CREATING YOUR Permaculture *Heaven*

How to design and create your own backyard food forest

NYDIA NEEDHAM

Copyright © 2021 by Nydia Needham

All rights reserved. No part of this publication may be reproduced or transmitted in any form or by any means, electronic or mechanical, including photocopying, recording, scanning or otherwise, or through any information browsing, storage or retrieval system, without permission in writing from the publisher.

ISBN: 978-1-915217-00-4 (paperback)

CONTENTS

Introduction vii

Chapter 1: An Introduction to Permaculture and Your Ecological Garden 1

Chapter 2: Permaculture Design—Getting Ready 16

Chapter 3: Permaculture Design—Design Principles 32

Chapter 4: Catching, Conserving and Using Water 52

Chapter 5: The Wonder of Soil 78

Chapter 6: Managing Expectations 106

Chapter 7: Choosing Plants for their Role in the Garden 125

Chapter 8: Attracting Birds, Bees and Insects to Your Garden 165

Chapter 9: Building Relationships in the Garden 184

Chapter 10: Garden Guilds 203

Chapter 11: The Food Forest 224

Chapter 12: Maintaining and Harvesting the Permaculture Garden 238

Chapter 13: Making Money from Your Garden 255

Conclusion 263

Appendix 267

Glossary 375

Bibliography 381

Resources 385

In the Action Plan for Permaculture Newbies, you'll learn:

— How to start thinking like a permaculturist
— An easy to follow mini guide to creating your permaculture heaven
— The 7 mistakes newbies make and how to avoid them
— Some mini projects to get you going

Scan the QR code below to be taken straight to your free copy!

INTRODUCTION

I remember the first time one of my friends helped me out in the garden. Everything was getting along fine, we had a lovely harvest of strawberries for later; then suddenly I heard her scream as if she'd been attacked by a lion. Well, there are no lions in my garden, I promise!

So I looked around. There she was, standing, really pale and scared. And on the ground in front of her was a life-threatening, dangerous, poisonous… earthworm!

I laughed. "Don't you know how useful those are?" I asked her. "They're a real gardener's friend; they improve the soil and help incorporate organic matter into it."

"But they're horrible!"

I sighed. It's not always easy to introduce people to permaculture.

A lot of us aren't brought up to get our hands dirty. We know about lawn mowing and decking and landscaping… not soil and worms and composting. We want a sanitized garden; we use pesticides and fertilizers and gas-guzzling ride-on lawnmowers to get them that way.

And 'big agriculture'—conventional, large-scale commercial farming—uses even more pesticides and more fertilizers and is actually pretty destructive of natural resources.

So imagine that you could use your garden to create a wildlife refuge for birds, butterflies and bees, to provide yourself with fresh fruit and vegetables and even craft materials for basketry and dyeing, to present you with a sun-trap for catching a few rays, a shaded area for when it's just too hot, the beautiful colors of a host of different flowers all year round… while also helping the planet?

Permaculture lets you do that.

It does so by basing itself on natural systems, just like one I've seen developing over the last few years. There's a little piece of woodland near us. Ten years ago, the owners cut down some of the trees for timber. They left the ground bare. The first year, there were grasses and a wonderful display of grassland flowers. The second year, a few shrubs began to grow, and now, there are a few trees just beginning to emerge from the shrub cover, but some of those pretty flowers have gone. On the other hand, there's a big blackberry bush that I have my eye on for making jam this fall…

Permaculture's basic idea is that if you look at these natural processes and try to copy them, you're going with the grain of nature instead of against it, and by doing so, you'll get a garden (or even a farm) that looks after itself.

Not everyone's a fan. Some people associate permaculture with gardens that look messy and wild. It's true, they don't have the classical appeal of a nicely trimmed lawn and neat herbaceous borders. But you *can* have an attractive garden and still use permaculture principles; it's all in the design. 'Permie' pioneers were after food, not flowers, but your priorities might be different. You can set up your own systems to provide the things that you want to get out of them, whether that's food, fuel, building materials, flowers for the house, or a home for a wide array of wildlife. In fact, if you want, you can have all of these things at once!

Some people think you have to have a lot of land to use permaculture. Inspiring pictures of forest farms—some guy who bought forty acres of land for a few bucks—don't help when you're starting with a tiny back yard, with a load of concrete left by the last owner, or a damp and overgrown garden with a lawn that ran wild and that has no shade and just claggy clay for soil. Actually, you can do permaculture on a balcony. Or you can start with that backyard and its poor conditions and transform it, bit by bit, into a productive and beautiful haven.

You don't have to be a shaggy hippy to adopt permaculture. You don't have to be a mountain man or an Earth mother. You can be a soccer mom or a fashionista (warning; you may get your nail varnish chipped) or a nerdy guy who works in software, or just a young parent who wants to do something for the planet and give the kids a place they can explore and learn, not just a patch of grass and a paddling pool.

You don't have to do everything. You can have a forest garden or just a single tree. If you have a balcony, you may not even have the tree. You can have a full-on wildlife garden, or you can run a 'normal' looking garden but with a permaculture philosophy.

Permaculture is a philosophy and a practice. So this book will talk a bit about the philosophy and also about the science behind it.

But I'm a practical kind of person and so this is also going to be a practical kind of a book. I'm going to show you how to go from concrete desert to fruitful forest or as far along that path as you want to go. One thing I want you to realize from the beginning is that none of this has to be daunting—you can start as small as you like and you can move at a pace that suits you. Don't feel you have to learn everything in one go and rush out to transform a whole wilderness in a year. Just deal in manageable chunks and you'll do just fine.

In this book, you'll learn:

- how to plan a garden that makes the best of the site, drainage and soil,
- how to improve the soil and make your plants much more productive,
- which plants to choose (by function and interaction with other plants),
- and best of all, how doing less work can get you more fruit, flowers and greenery.

I grew up in the country and that had a couple of different impacts on my life. First of all, I learned to love nature, to enjoy the great outdoors, hiking or watching birds, or just watching the clouds sail past. But I also learned how modern agriculture can destroy a river's natural life by pesticides leaching into it from the soil. I was shocked just how much artificial fertilizer the prettiest garden in our village used, despite the fact that there were stables producing loads of manure just down the road.

I got interested in ecological agriculture and the green movement, one thing led to another and I ended up learning about permaculture as a way of creating a sustainable system of food production. But it wasn't easy—there's a lot of science to learn. Some of the early writers on permaculture are ninety percent science and only ten percent practical gardening. In this book, I want to do things the other way round, only giving you the science you absolutely need—and I'll explain it very simply—but sharing a lot of advice on practical issues like designing your garden, choosing plants and improving your soil for better returns.

On the practical side, I started out myself by working on an allotment with a friend. The plot was overgrown, but it did have a couple of apple trees that were still bearing fruit. We both did a lot of research into permaculture, reading the early works of Bill Mollison and David Holmgren and then moving on to more recent information from the likes of Geoff Lawton, Toby Hemenway and Patrick Whitefield. We got to know about as much of the science as we could before moving onto the practical aspects and getting our hands dirty. Before this, we had both grown fruit and veggies in our small backyards, but this was a whole new adventure. We started by sheet mulching the whole plot so that we could start from scratch. We built a few raised beds in the meantime so that we could still enjoy our favorite fruit and veg. To this day, raspberries are the crop that provide me with the most pleasure, being one of my favorite fruits, so they had to be in there right from the start! We thought about our water supply, built a few swales and had a few tanks to catch rainwater (there is definitely a lot of that in the UK). Since then, we slowly built up the allotment with more trees, perennial fruits, veggies and flowers, a few hedges, and around these, we plant our favorite annuals each year. Yes, we've made mistakes and it hasn't always been easy, but we have learned from them and carried on, and are still learning new things every day! Now I want to share what I've learned, both from researching the theory and actually putting things into practice.

Whether you have a small backyard or an extensive ranch, or just a balcony, you can do permaculture, and the principles are the same, though you'll obviously need to apply them differently. There's an exercise at the end of each chapter to help you through the process.

But because permaculture is a design process, a philosophy, an art and a science, there's a fair amount of reading to do before you get your hands dirty. So don't rush out and buy yourself a load of tools or plants before you've read the first few chapters and understand what it's all about. At the same time, once you've understood the basic principles, remember that you can tackle the practical stuff a bit at a time, and then this book can become more of a reference guide that you can come back to whenever you need it.

Most gardeners (Geoff Lawton, who's done permaculture all over the world, is an exception) and most gardening books only cover one climate. The 'fathers' of permaculture, Mollison and Holmgren, were Australian and their books are based to a large extent on Australian conditions. I'm based in the UK and of course, my knowledge of permaculture is based on our climate, which for American readers is USDA 8-10, mainly sitting around zone 9. Winters don't freeze really hard and summers usually aren't all that dry. But I've done quite a bit of research to try to find options for everyone, including some sub-tropical and tropical guilds. And wherever you are, the principles of permaculture are the same, so feel free to experiment along those principles with local plants.

Let's get started!

1

AN INTRODUCTION TO PERMACULTURE AND YOUR ECOLOGICAL GARDEN

In this chapter we won't be getting our hands dirty. We'll be looking at some history, some philosophy, a little bit of science, and trying to answer two basic questions: What does permaculture mean? And why is it so important for the planet?

If we look back in history, permaculture isn't a new idea. Indigenous peoples in the Americas were using it long before Columbus arrived. Archaeologists and botanists have found traces of civilization in the Amazon, showing that the apparently natural landscape was being managed to produce food. Thousands of years afterwards, we can still tell where the Mayan and other indigenous cultures grew their gardens in the forest.

In the southwest US, native peoples practiced cultural burning, using controlled burns of forest to create glades and meadows where smaller shrubs had room to grow, giving them materials for basket and house making and medicines, as well as berries for food.

In the Himalayas, there are gardens using what we would call permaculture principles. The southern state of Kerala, in India, has similar forest gardens, combining coconut palms with shrub and salad plants. Indigenous farming systems often concentrate on enriching the soil—for instance, in the Himalayas, terracing retains the soil, and rotted down human waste is used as fertilizer (every Ladakhi house has two compost toilets, one for this year and one for next year). After a few centuries of this farming, the soil is black and rich and the crops abundant.

These people understood nature and created polycultures, that is, agricultural systems where different plants share the space, instead of modern monocultures like big wheat fields without hedgerows.

Permaculture works in any climate. It's not a discipline, as such, but a design approach that uses different techniques all based on the idea of natural systems and the way they operate as a whole, dynamic and alive, plants and light and soil and water all interacting with each other. You can look at it as a philosophy, or a toolbox, or a methodology—whichever way works for you. The important thing is not to lose sight of the basic principles.

Modern permaculture has its origins in the work of two men, Bill Mollison and David Holmgren, in the 1970s. They wrote two of the basic texts on permaculture and founded Permaculture Institutes around the world. Mollison worked in Brazil, Hawaii, Fiji, on city farms in the UK and US, and advised on arid land techniques in Zimbabwe and Vietnam; permaculture can go anywhere!

Bill Mollison started off as a jack of all trades—baker, shark fisherman, naturalist, tractor driver, trapper, mill worker, museum curator. But his career really got started when he started working for the Commonwealth Scientific and Industrial Research Organisation in the Wildlife Survey Section, and then when he decided to sign up for a degree in bio-geography at the University of Tasmania. Among other things, at CSIRO, he had to deal with plagues of locusts, the problem of (non-native and rapidly multiplying) rabbits, and fisheries research.

He was amazed by the abundance of the Tasmanian rainforest and decided to try to create systems that would function in the same way, to enable sustainable agriculture to replace the industrial growing methods that he saw impoverishing the soil and poisoning the water in Tasmania. Working with David Holmgren, he created the term 'permaculture' for a framework of intercropping perennial trees, shrubs, herbs and fungi.

Where Bill Mollison was revolutionary was not in his understanding of the natural forest systems—scientists had already been researching those for some time. He wasn't revolutionary in seeing Big Agriculture as ecologically damaging—Rachel Carson's *Silent Spring* had already pointed this out in 1962. His great leap forward was in believing that we could create a similar system to the rainforest in our own gardens and agriculture—a self-sustaining, fertile, productive polyculture. He created a village, Tagari, where his ideas were applied—so what he knew in theory, he put into practice.

His vision was of permaculture as a 'living clockwork,' a perennial system, self-sustaining and biodiverse, which minimized input (for instance, human work or fertilizers) and maximized yield. He believed it could be applied by anyone, on however large or small a property. And he believed that diversity was linked to stability—the more plants and the more relationships between those plants, the more stable the system would become.

For Mollison, permaculture was a return to the age of innocence. "In short, a permaculture should be nothing less than a Garden of Eden," he said.

While Mollison travelled the world teaching and promoting permaculture, Holmgren concentrated on refining both his mother's property in southern New South Wales and his own property, Hepburn Permaculture Gardens, Victoria.

Since 1991 he has been offering Permaculture design courses and from 1993 has been teaching from his Hepburn home.

A slightly different development happened in England with the work of Robert Hart. He took over a small holding in Shropshire but discovered that he simply didn't have the strength to manage a traditional farm with the need for tilling, digging, pruning and annual sowing. Instead, he started looking around him for inspiration in nature (he also received some inspiration from the work of Japanese social reformer Toyohiko Kagawa, who pioneered the planting of trees by upland farmers to save the soil and provide fodder for animals).

One of the most beautiful sights in England is an old hedge, with maybe a couple of ancient oak trees along its length. You may look and just see 'a hedge', but in fact, it will be made up of various species. You can even tell the age of a hedge by the number of species in it; the older it grows, the more diverse it becomes. There will probably be blackthorn or hawthorn, with spikes intended to keep animals out of the arable fields; there might be hazel, wild plum, crab apple, and brambles will probably snake their way between the branches so that by September, you'll have blackberries to make into jam or jelly.

There will be undergrowth plants, maybe cowslips or wild garlic on the bank under the hedge, and there will be insects and particularly bees visiting the flowers, birds making their nests, rodents living in little holes between the roots. Probably there will be moss and lichen on some of the older growth. It isn't just a closure, like a barbed-wire fence—it's a whole little world that creates a protected environment for many species to live in.

And that's the kind of dynamic, living system that we're trying to recreate when we do permaculture gardening.

But a lot of English village gardens looked completely different, with straight lines of cabbages marching with military precision across the bare soil of the vegetable beds. There were tomato plants that needed watering twice a day, morning and evening, and also needed regular doses of 'tomato food,' which came out of a packet and probably included quite a few ingredients that are banned today.

When the wind got going, there would be a little dust devil in the garden, taking the top layer of unprotected soil away.

Hart decided his garden should have more in common with the hedgerow, a garden on different levels, from the tallest tree down to the ground-covering plants at the bottom. He created the forest garden on his tiny patch of hillside, and English permaculturists like Patrick Whitefield followed his example.

Don't confuse permaculture with a wildlife garden. There are similar ideas behind them, but a wildlife garden is intended to serve wildlife by providing frogs, hedgehogs, birds and insects with the habitat they need. A permaculture garden is also intended to serve its human users by producing a food harvest for them.

Don't confuse permaculture with 'letting things go.' True, many permaculturists have gardens that look, to be honest, a bit untidy, but that doesn't have to be the case. Permaculture will never suit minimalists, but how tidy your permaculture looks is entirely up to you. You can have a neat, tidy permaculture garden or a rambling, sprawling one, but the ideas behind it will still be the same—a natural self-sustaining system that doesn't need fertilizers and pesticides, that can look beautiful and also feed you and your family.

As well as food, you can produce lumber, edible plants, herbs, medicinal plants, and craft materials like wicker for basketry, greenwood for spoon making or bowl-turning, or vegetable dyes. And while permaculture aims to make a difference by helping you become more self-reliant, there is no prohibition on having some flowers that you like because they're pretty—though they should, ideally, fulfill other functions too (like providing pollen for bees).

Mollison's original word 'permaculture' is a sort of mash-up of 'permanent culture' and 'permanent agriculture.' He saw his permacultures as integrated designs—the garden or farm designed as a system, just like an IT system, in which everything is related. Systems thinking looks at relationships, not just things. When most people look at a picture of a plant in a garden catalog, they think of it as a single plant—"oh what a pretty rose," "I like the idea of growing my own cherries." But when a permaculturist looks at the picture, they will be thinking, "that rose is a rambler, so it could climb up a tree, and it could provide shade for the plants underneath, and there will be rose hips for the birds in autumn, so the birds will manure what's underneath and they'll eat pests."

Permaculture integrates different scientific fields of knowledge; hydrology—knowing how water systems work—and soil science, biology and meteorology, geology, even animal husbandry can all come into play in a permaculture garden. Modern agriculture, on the other hand, keeps knowledge in separate 'silos,' and like the typical gardener, looks at a crop as a monoculture that doesn't relate to any other plants or life forms. It doesn't try to feed the soil except with fertilizer (which is like trying to live on a diet of vitamin supplements instead of eating fruits and vegetables) and it uses huge amounts of energy and chemicals in the pursuit of higher yields (then even more energy gets used to transport the food to a different country, to different supermarkets, and from the supermarket to you. Inevitably, most of that energy will use fossil fuels and contribute to global warming).

Intelligent design saves maintenance. That's true in engineering—everyone knows there are some models of cars that never break down and others that spend all their time with the mechanic—and it's true in permaculture.

Maybe permaculture isn't going to take over the world just yet. And maybe your little permaculture garden won't provide you with all the food you need. But every little step you take to make your backyard (or farm, or balcony) more productive will help swing the balance, from humanity being a destructive force to being a positive one.

So let's take a look at a natural system and how it works. A basic principle is that nothing grows forever. There's a cycle of growth, decay, and rebirth that continues. The seed grows, the crop is harvested, the leaves fall, they rot down and fertilize the soil, where another seed is sprouting. So a permaculture garden is a dynamic thing. It's the opposite of a Zen garden where the rocks never change—it's *always* changing. It changes with the seasons and it changes with the years. Your aim is to work with the cycles of nature, and by doing so, you're using nature's energy as well as your own to achieve the same objective.

In fact, by using our intelligence, we can help the ecosystem mature faster than it would in nature. And of course, we can take a few shortcuts, like buying fruit trees that are already four or five years old instead of growing them from seedlings.

Beneath the growth-decay-rebirth cycle, there are deeper cycles, bio-chemical and geochemical cycles. Carbon, oxygen, sulfur, nitrogen and phosphorus are used or emitted; trees breathe in carbon dioxide and some leguminous trees fix nitrogen in the ground. So besides considering your plants, you need to be thinking about the soil and the elements and microbes and fungi that it contains. A garden is never just plants—there are always bacteria working away, insects pollinating the flowers, microbes, fungi that attach themselves to the plants' roots. Like an iceberg, any garden is ninety percent underground—it just that in permaculture, we understand that, while other gardeners only look at the bits they can see!

With all these cycles, permaculture is about trying not to lose energy or nutrients, or resources. For instance, you'll learn how to manage water so that rain doesn't wash away the soil but is stored and used when it's needed; green manures are chopped down and left to make a mulch that will rot down and provide nutrients to the soil. The cycle shouldn't 'leak.'

Within this system, each species depends on one or two key factors in its relationship with its environment. To ensure it thrives, you need to know what those are.

For instance, some seeds need to experience a period of intense cold before they will germinate—so they need an environment that has a hard winter. Some plants need a certain number of hours of direct sunlight; others need shade. Some plants need birds to help scatter their seeds; others need a taller plant or structure up which they can climb. Other factors may also affect its growth but won't be crucial.

The system also needs to be diverse. If a population, or a species, is too scattered, it's likely to become extinct. You may have seen this if you've planted three or four daffodils in a sizable garden—eventually, they just give up, whereas if you plant a good size clump, they'll probably do well. But—and this comes as a surprise to most people—when a population is too dense, it's also likely to die off. That's why, for instance, irises need to be divided every so often.

Diversity isn't just about having, say, one hundred varieties of plant in your garden. It's about the relationships between those plants. How can we build those relationships so that the plants help each other? Permaculture is a vision of a cooperative ecosystem. You're looking at your plants as if they're a football team—you need a coach, you need an offensive unit, you need linebackers and place kickers and you need medics and physios as well. If you just have eleven place kickers in your team, you're not going to win many games.

Mollison and Holmgren used these basic principles to identify the factors that made the forest systems they had studied so abundant and so sustainable. They had a vision of systems that contained integrated, self-perpetuating populations of plant and animal species, systems that would require little or no human intervention. By copying natural systems, they could combine low maintenance with high yields; and they could achieve stability, diversity, *and* energy efficiency, all at the same time.

A big principle in permaculture is that of redundancy. If you work in IT or telecoms, you probably know about how redundancy can create resilience. In permaculture, it works in two ways;

- every feature in the garden should fulfill more than one purpose, and
- every need should be filled by more than one feature.

So, for instance, a plant should do more than one task. A hazel bush can provide you with nuts to eat, but it can also provide you with sticks for making trellises and supports for beans and other climbing plants; the pollen from the catkins attracts bees and other pollinating insects; and hazel can also be used in a hedge, as a windbreak or for your privacy.

Look at a fruit bush in a natural forest. It has higher trees protecting it from the sun and smaller plants protecting the soil around it from moisture evaporation. Put it all alone in a bare earth border and it's got much more work to do because it's alone, not helped by the other members of the team. And you have more work to do because you'll need to water it all the time. So when you're designing your garden, don't think about single plants—think about the structure and the system.

If you keep ducks, they will lay eggs; they will also eat slugs and weeds and pests. It's up to you whether their other benefit is having amusing, charming pets or having a supply of fresh duck meat.

Let's look at some of the problems of 'normal' gardening. A huge problem is that there are often large areas of bare Earth. In nature, what happens when the Earth is bare? Pioneer plants come along, see an opportunity with no competition, and grow fast and wild. You probably don't call them pioneers, but 'weeds.' They've evolved to make use of this specific kind of opportunity to spread and seed themselves fast. So if you have bare soil in your borders or your vegetable garden, you're going to spend half your life weeding.

Your plants don't form relationships and so they never create a stable system. In a forest, for instance, the trees throw shade and stop the shrubs underneath from growing too big. And the pioneer plants can only grow on the edges because they don't get enough light and there's too much competition in the middle of the woods. But with a lawn, for instance, you have an environment that hasn't matured. What you're doing mowing the lawn is the equivalent of trying to stop a baby from growing up. So again, you're spending time doing maintenance and mowing the lawn every weekend. Note however that in some parts of the world, mowing may be necessary to prevent wild fires from spreading.

Bare soil also lets moisture evaporate, so you need to water the plants. And when you clear up all the fallen leaves, and twigs, and dead plants, and weeds, you're taking away all the organic matter that could have put goodness back into the soil. If you take it to the compost heap, it might rot a bit faster—but what a waste of energy taking it to the compost heap just to bring it back a year later and put it on the same bit of garden! Why didn't you just leave it there?

So permaculture can help you do less work, at the same time as making your garden more productive. And as a permaculture garden matures, there will be less and less work to do. In fact, most of the work you'll be doing is picking food and eating it.

There are three principles in permaculture; care for the Earth, care for people, and fair shares.

Care for the Earth is about caring for precious resources. Sometimes the full impact of your actions isn't obvious till you think about it. For instance, storing rainwater in a cistern doesn't just mean you get 'free' irrigation; it also helps reduce soil runoff, it reduces the amount of water going into the sewage system, helping prevent it from becoming overloaded, and potentially helps prevent flooding in the waterways.

Planting an apple tree can stabilize the soil on a hillside, as its roots spread into the Earth; its fallen leaves will gradually be absorbed into the soil, giving it nutrients and it will do its job of taking carbon dioxide out of the air. And every apple you don't have to buy from the supermarket is a saving in transport cost and your personal carbon footprint.

(Mollison asserts what he calls "the life ethic," that every living thing, whether plant, bacteria, microbe, or animal, has a value, independent of its usefulness to us).

Care for people starts with ourselves and taking responsibility for our own existence. Permaculture means stepping away from consumerism, but it also means eating more healthily by eating organic food that you grew yourself. It can mean greater enjoyment; if you've never cracked a pod of peas open and eaten them fresh, or bitten into a juicy, crunchy apple right off the tree, believe me, you haven't lived!

Fair shares doesn't just include sharing with other people—though many permaculture gardeners share with their neighbors, with their friends, and even with local food banks. It includes sharing with wild animals or perhaps with ducks and chickens. It's about having abundance that enables us to share without going short ourselves. It's about reinvesting—for instance, through installing a greywater harvesting scheme or composting. And it's about sharing our labor, our expertise and our knowledge.

A phrase you'll hear a lot is "Each one teach one." Fair shares includes being generous with your knowledge, and it's a great way to achieve those other two principles since anyone you've taught to care for the Earth and to care for people is going to be carrying on the same great work that you're already doing.

Let's get back to Mollison. He saw how destructive modern agriculture had become. That was particularly true when it came to the topsoil; by plowing deeply every year and by leaving bare Earth between plants, agriculture was exposing the topsoil to wind erosion. Irrigation carried away even more topsoil, and intensive agriculture depleted the nutrients in the soil and put nothing back.

What he observed in the forest was the way different plants and animals related to each other; he saw how nature is a network of relationships, and his vision is that if you design a garden with that network of relationships in mind, you have a permaculture that works with nature and not against it.

So when you do permaculture, you don't do a whole load of digging, rotavating, planting. You start off by *thinking* because you are designing a system. It's like designing a clock; you need to put different cogs in different places, you need a source of energy, whether it's a spring or a battery, you need a way to display the time… and you need to get all of that machinery put into a certain finite space.

Modern agriculture sees land and plants as things that are basically simple and not related to any other parts of the system. Permaculture practitioners look at the land first, then design around it, trying to maximize its potential. If you think of the way great French vineyards are all designed around a certain orientation and degree of slope to get the maximum sun during the growing season—that's a permaculture approach. Traditionally, they also used to put a small peach tree at the end of each row of vines to attract leaf blight. It was like the canary in the mine—if the peach tree got blight, you had enough time to treat the vines before they got the disease, and you'd saved your harvest (and if the peach tree didn't get blight, of course, you had tasty peaches to eat). That's a permaculture approach, too—companion planting, using the relationships between different plants purposefully.

You might think that permaculture is a bit of a new-age fad, like dreamcatchers or crystals or doing yoga. The statistics suggest otherwise. Mollison noted that when China adopted modern agriculture, farm energy use rose by over 800 percent, but farm yields rose only 15% (would you do nine times more work for just a 15% raise in salary?).

In other places, modern agriculture practically ate itself, destroying the environment on which it depended. The Californian orchards took so much water from the aquifers; they've been suffering drought ever since. The great Dust Bowl, a huge disaster for American farmers, came about because instead of understanding the Great Plains ecology and its needs, farmers deep-plowed the soil. By the end of the 1930s, it's estimated some areas had lost more than 70 percent of their topsoil—blown away by the wind.

That's how destructive 'modern agriculture' can be. And the worst of it is that all that deep plowing used up human energy, as well as fossil fuels, just to achieve such a disastrous result.

A permaculture designer would have looked at the Great Plains and thought about its ecology before taking a decision. Modern agriculture, Mollison said, demanded that the land gave it what it wanted regardless of its ecology. Instead, permaculture asks the land "what can you give me"?

Let's take a global view. The Earth, with its current population, has a little under 5 acres of land to meet the needs of every person alive. However, according to a study from the University of Wisconsin, suburban American lifestyles need more than 31 acres per person. That includes farms, roads, orchards, forests, and their houses, as well as distribution and retail facilities. Even just going without meat one day a week could make a huge difference.

Modern agriculture is one of the biggest emitters of carbon dioxide; combined with forestry, it's responsible for up to a third of all emissions (those figures come from the Consultative Group on International Agricultural Research (CGIAR), a partnership of 15 research centers around the world).

Carbon is emitted by tractors and combine harvesters, by heating for greenhouses and electricity needed for pumps and irrigation. Farms are responsible for the emissions of the chemicals industry when it makes fertilizers, for the extractive industries that mine phosphates for fertilizer, for the emissions involved in transporting that fertilizer to the point of use. Cooling down a big chicken shed takes energy.

So a reduction in the carbon resources used by agriculture could make a big impact on greenhouse gas emissions and help address global warming. On a tiny scale, consider how if you set up a wormery that uses your kitchen scraps to fertilize your balcony garden, you've replaced a whole energy-sapping lifecycle of fertilizer production, transport, and even the plastic wrapper for the fertilizer.

Being self-sufficient, or at least that little bit more self-sufficient than we were before, helps the planet. If you're producing some of your own food, less land is needed for factory farms. Less damage is being done to the environment from pesticides, fertilizers, deforestation, water extraction, or soil depletion. You may be seed saving and passing seeds for heritage species on to other gardeners, instead of using sterile hybrid seeds or genetically modified variants. And you will also acquire skills and knowledge you can pass on to your family and friends.

So permaculture's not just about 'self-sufficiency,' like an isolated hermit living off-grid—it's about creating a more sustainable society and economy for all of us.

If you grow all your own food, that's great. If you have enough left over to contribute to a local food bank and share with friends, that's fantastic. If you just supply your household with fresh tomatoes for three months of the year, you're still saving a huge amount of carbon compared to buying tomatoes produced in a hydroponic growing system, delivered by air cargo and then through road transport to a supermarket. You will also be doing the following things;

- Putting back nutrients into the soil.
- Reducing pesticide and fertilizer use.
- Providing food for pollinating insects.
- Shifting away from a consumption economy to produce your own food.
- Using renewable resources.
- Reducing waste.
- Helping to spread the message about permaculture and responsible farming.

Even if you live in an apartment, you can make your balcony into a permaculture garden using the same principles that apply to a smallholding or an agroforestry project.

Permaculture can be restorative. Across the world, many areas now have big problems after years of intensive agriculture have stripped the goodness out of the soil, extracted too much water and damaged biodiversity. For instance, in California, the fruit orchards have been extensively irrigated for over a century; now, as farmers dig deeper and deeper to find the aquifers, the water coming up is increasingly salty. The Sustainable Groundwater Management Act passed in 2015 aims to reverse that process but doesn't see the aquifer being sustainable (that is, farmers taking out less than flows in) before 2040.

Regenerative agriculture aims to put resources back into the soil and the land—improving soil quality, putting nutrients back into the soil, stopping runoff and soil erosion and even using the soil as a carbon sink to remove excess carbon from the atmosphere. A regenerative system can improve water retention and increase biodiversity. It's holistic, working with the entire ecosystem, not just isolated parts of it, and the aim isn't just the one of preventing further harm to the land, but of reversing the damage and making things better.

So permaculture is ethical (and some people may even see it as political) as well as being a philosophy of using natural systems.

Let's take a look at what permaculture *isn't*. It isn't a cult. You don't have to do everything it says; there is no 'holy writ,' and while some permaculture practitioners may find your decision to have lots of black colored plants in your Goth or Emo style garden a bit odd, as long as you stick to the basic ideas of permaculture, your taste is up to you alone.

It isn't a fan club. Lots of permaculture gardeners enjoy networking with each other, but you don't have to. You can carry on playing golf and have a permaculture garden. You can carry on playing Dungeons & Dragons and have a permaculture garden. You don't have to join any clubs or societies.

It's not exclusive. You can have a forest garden, a balcony garden, a wildlife garden. You can use other tools and techniques. You can have a permaculture back garden and a lawn in front. You can have formal parterres surrounded by wicker fences and a neat orchard planted in lines, but your planting within the parterres will use permaculture principles to get the best out of the vegetables, and your apple trees will have daffodils planted in a circle around them to stop grass from competing with the tree for nutrients.

Or you can have a great big untidy-looking Hugelkultur bed made of twigs and mulch and a huge harvest of vegetables. It's all up to you how you balance these elements.

It's not 'gardening.' In fact, it's 'anti-gardening,' in that some permaculture gardens are completely no-dig, no-prune, almost zero work. You could call it 'designing with nature' or 'channeling nature.'

Holmgren's Twelve Principles

1. **Observe and interact.** Look at the site in all seasons, design for a specific site and culture. Link up the parts, build relationships and connections.
2. **Catch and store** energy and materials. A garden is a series of flows—every cycle can produce a yield. Reinvest resources.
3. A garden is an **investment** on which you get a yield. Set up feedback loops to repay you.
4. **Self-regulation**—accept feedback. If you don't need to do something, don't do it! If something needs to be done, do it.
5. Use and value **renewable** resources.
6. **Zero waste!**
7. **'From patterns to details'**—look at patterns in nature and make these the backbone of your design.
8. **Integrate, don't segregate.** This is about connections and relationships again.
9. **Small and slow is better than big and fast.** Use local resources, start with a small pilot plot, produce sustainable outcomes, reinvest to build on success and repeat what works.
10. Use and value **diversity**. Let diversity create resilience.
11. Value the marginal. **Use the edges**. Intersections are the most diverse places and they're the places where everything happens—the riverbank, the hedgerow, shallow water.
12. **Use change creatively**. Observe, and intervene at the right time.

It's not a template. Although we're going to talk about 'guilds' later, collections of plants that work together, you can't cut-and-paste permaculture. You need to look at the site, the soil, your own taste (both visually and in food) and design what works for that particular combination. It's not a 'kit'; because you're not just sticking things together like LEGO, you're creating relationships between them.

There's another thing that permaculture definitely isn't. It's not a fashion or a fad or a craze. It's been going since the 1970s and it reflects cultures that go back much, much earlier than that.

CREATING YOUR PERMACULTURE HEAVEN 15

So let's take a look at some ways you might create a garden with permaculture. You might create a food forest, reproducing the way the Amazon rainforest or a temperate forest grows on different levels—the canopy, smaller trees, shrubs, undergrowth. Your food forest might just be one tree if you have a small garden, but instead of standing in splendid solitude, it will be at the center of its own little ecosystem.

If you don't have room for trees, you could still use this concept for vertical planting, which is a very efficient way of using a small space. Plant stacking lets you grow vines, gourds, or passion fruit, up a trellis or over a pergola; then underneath, you can have your shrubs and then smaller plants like salads or root vegetables.

You can stack plants in time, too, with succession planting. So, for instance, you sow radishes and early salad at the same time. The radishes will grow fast, and so when you pull them up to eat them, you're automatically thinning out the salad plants with no waste, compared to the usual gardener who just plants lettuces and then thins out and throws away half the seedlings.

If you have a larger space, you can create tiny microclimates. Putting trees in a 'U' facing south can create a sun trap—great for catching some rays or for growing melons and other warmth-loving plants. A pond can help reflect light into your greenhouse and raise the temperature a couple of degrees. Or you could create a water garden using a reedbed to purify your greywater (washing up water, for instance) before it goes into a pond with ducks and edible water chestnut, and then to other areas of the garden that need it.

Almost incredibly, a regular suburban backyard can be three times more productive than farmland (reference: The Garden Controversy: A Critical Analysis of the Evidence and Arguments Relating of the Production of Food from Gardens and Farmland—Robin Hewitson Best, J. T. Ward, University of London). If that's not a great argument for permaculture, I don't know what is!

2

PERMACULTURE DESIGN—GETTING READY

The important thing about permaculture is that it's not haphazard—it's not something you can do just by picking a few favorite plants or choosing a water feature—it's a system. And because it's a system, just like an IT system or a production system, you have to design it first. You can't get your hands dirty before you've done the hard thinking!

Some of you will read that and feel quite happy that there's an intellectual element to permaculture, and you'll be keen to engage with the system design process. Others may be impatient to get on with the digging. You bought a book on gardening and you've got to read a load of theory?

Well, you do need some theory because it will help you be smart with design, so you'll end up with a garden that needs a lot less work in the long run. It's really worth taking a little time for that, and you can think of it as a little investment now for a big pay-out later. You are designing an ecology. You are starting a process of natural evolution. You are designing a little world. And even God took seven days to do that; reading a few chapters on permaculture design won't take *that* long! Do remember too that you can return to this book as a reference guide, so don't worry if you don't take in and memorize every last detail on the first pass!

THE BASIC STEPS

Observation is the first step. You start by looking and noticing things. Look at the place you're going to put your permaculture garden or farm and work out where the water flows, where the sun shines the most, where the ground is hard or boggy.

Mapping is the most important way of recording your observations. And once you've done that, you can use your map to help the next stages in the process, too, such as working out where things ought to go. As part of mapping, you'll want to look at:

- zones—where *you* want things to be, and
- sectors—energies coming into your garden (wind, water, etc.).

Envisioning is the next stage. You need to sit down with all that you've learned about the site and you can start to envision your garden—to get an overall feeling for what you're trying to achieve, whether that's a vegetable and flower garden embodying permaculture principles or a full-scale food forest, and how that fits with what's already there.

Think about your focus—how you want your garden to look and feel. Will it be a wildlife refuge, an orchard producing fresh fruit, a cool, relaxing place, or your workplace?

Planning naturally follows the envisioning. You can begin to think and record more formally about where water courses and plants ought to go.

Development/design takes you a little further into the specifics, and in particular, thinking about how you are going to organize the work and how long you're going to take and thinking about the progression of your garden over time.

Implementation comes last. This is where you can start getting your hands dirty at the end of the process.

While you're doing all this, you need to keep those principles of permaculture we talked about in the introduction in mind. So, for instance, everything needs to have more than one function. Your vegetable garden fence could keep chickens out, be a windbreak, and provide shade for plants that need it, all at the same time.

Now let's look at these stages in more detail.

OBSERVATION—LOOKING AT YOUR SITE

So, where can you create your permaculture garden? You may have the idea that you need to have a food forest, a few hundred acres of wilderness… well, that works. But you can adopt permaculture principles in your backyard, in a balcony or even a window-box, in an allotment or shared garden, or, if you live in an apartment block, perhaps even on the roof!!

Garden shares are becoming increasingly popular; people who, for whatever reason, can't cope with a large garden are sharing with keen gardeners who don't have enough land. Using a friend's yard to establish one permaculture 'guild' (a cooperative collection of plants—we'll talk about these later) could be a good option for some. In fact, even if you're planning to buy a larger plot in the backwoods and make your living from permaculture, you should definitely put in a couple of years on a shared garden (or go and work on a permaculture farm) first.

And why do permaculture? After all, you could have a big lawn like everybody else and spend your weekends mowing it and then putting on the sprinkler. There are a whole load of good reasons to do it, however small your garden, because every little bit you do makes a difference. You're increasing biodiversity, even if you're just doing it by saving an old tomato variety or getting birds coming to your balcony for berries in winter. You're creating habitat for bees and other pollinators. In Paris, there are even hives on top of Notre Dame Cathedral, and yes, they survived the fire. You're reducing outside inputs and reducing transport costs. If you grow sweet potatoes, they come to you from a hundred-yard walk down the garden, not a thousand-mile trip by airplane. You're helping to enhance the quality of the air, the water, and the soil. You're reducing waste. You're helping wildlife. What's not to like?

Or maybe you're just busy or lazy and want a garden that will look after itself—since the idea of permaculture is that you're setting up a natural system that should maintain itself with the minimum of human intervention—although to be honest, very few permaculturists are lazy; they get the bug and want to do more.

So, where do you start? If you only have a window box, well that's fairly easy; grab your tomatoes and herbs and some companion plants, some compost, and get started. You'll have fresh tomatoes and herbs, it'll make you feel good, and you will have done your tiny little bit for the planet. Get a wormery, feed the worms your kitchen scraps, recycle those into compost and you have the fertility to feed back into the soil for your plants.

But for most spaces, you need to start your process of creating a permaculture garden by looking methodically at the site. For permaculture to work properly, the right stuff needs to go in the right place at the right time. So the first thing you do is map the site. In particular, you want to know:

- Where's the water?
- Where's the access?
- Where are the structures?
- What is the slope?
- Where are the contour lines?
- What's the soil like?

Getting the water flow right is absolutely basic to permaculture. Try to get your water reserves as high as you can so that gravity delivers it to your garden automatically. Find the highest and longest contour lines. Usually, if you have a boundary along the highest side of your property, the lowest point of this highest boundary will be a place where water collects; you might dam it there, for instance, to make a reserve pond. If your house is at the top of the slope, rainwater from your roof can be collected in a tank; if your house is nearer the middle, you might want to raise the tanks off the ground to give you some head pressure.

After that, you'll be thinking about how to spread the water around the plot and soak it in so that all your soil is well irrigated. If you just leave things to nature, water will flow straight downhill, eroding soil and creating gullies, so your aim is to slow it down, to store it (if there's a place that's already hollow or boggy, that could become your storage point), and to let it gradually penetrate the soil and water your plants.

The easier you make watering, of course, the easier your life becomes. If you've ever caught yourself saying, "I want to go away for a few days, but who will water my tomatoes?" then you need permaculture in your life!

Think about how to harmonize access with the water pattern and other aspects of the site. For instance, can you use a road as a firebreak or as a water-catchment area? Think about where things are in relation to each other. For instance, if you're using your home garden, then the position of your house is a given, so you need to think about how you reach the compost bin from the kitchen, which walls of your house are warm (south-facing, for instance), whether you have areas that are inaccessible like a side alley.

There's a lot to think about: water flows, geology, organisms, birds, spiders... You need to see the details but also see how they flow together into a whole and how they connect. Some gardeners describe the right state of mind as almost a child-like state of wonder—"How do these things all happen? How do they all fit together?"

Don't rush your observations; different seasons, different times of day or weather, will all show different aspects of the site. Some permaculture practitioners don't like to make any decisions till they've lived a full year or more with the land, observing a full cycle of seasons. If you go out in heavy rain or wind, it can be really informative—where does the water erode your soil? Which areas are sheltered from the wind, and which are exposed? Go out very early in the morning in spring to find out where the frost pockets are.

Which animals are in the garden? Where do they live? Where do they go and why? What do they eat? How do they interact with other animals, with plants, with humans? For instance, your pet cats may have created their own paths through the garden; they interact with you for petting and food; they chase and catch mice and birds; they don't interact with the chickens; they eat grass occasionally.

Notice, too, if there is any pollution coming from roads, factories, or industrial farms.

MAPPING

Mapping is how you record your observations. You can get started in various different ways: get a base map from Google Earth, or a contour map, or draw your own map. Actually, drawing your own really helps. A lot of maps show things you don't care about, like municipal boundaries and house names, and don't show things you really care about, like which way a river is flowing.

Make a base map that has the really immovable elements—the watercourses, roads, walls—that are already there. Then keep a clean copy of the base map (on paper or digitally) and use a separate copy to mark up what you observe. These should include:

- Structure—sheds, pools, earthworks, fences.
- Plants and trees.
  - Note them as DAFOR—dominant, abundant, frequent, occasional, or rare.
- Functional equipment—anything that performs a function, such as a water pump, a solar lamp, a heater for the greenhouse. You might already have put some of these under structures.
- Animals, insects and birds. You may see tracks (spoor or 'desire paths') worn by the frequent passing of

animals. Ask yourself where they are going and why. You might notice birds like particular plants or berries.
- Sectors—that is, energy coming into the site. Sun and wind are directional and can be shown by an arrow. Note any topographical sectors where the landscape creates frost pockets or flooding areas, wind tunnels or shade, soggy or muddy areas, or erosion.
- Look for microclimates! Where is sunny, even on a cool day? Where is cold and damp? These areas will be really useful when you want to add plants with specific needs, such as plenty of sun or moisture.

You should also test soil across any larger site. Feel it—is it sandy, crumbly, or clingy clay? Different areas may have different soils. You can use a jar to do a settlement test; mix a handful of the soil up with a lot of water and sand will settle at the bottom, then silt, then clay (you need to take the soil for the test from a few inches down, not just the top layer). Now you know exactly what you've got. Put that on the map too.

You may think drawing maps is a waste of time, but it will save you from designing a garden that doesn't work, wasting time and money, and then having to put things right.

You could draw each factor on a separate map or on a separate layer of transparent plastic so you can overlay each map on the others. The types of map you may need include:

- Base map—just showing the boundaries, any buildings and roads, maybe from Google Maps or another public source.
- Water map—spigots or springs, downspouts, French drains, swales, rivers and streams, containers, pools; remember every roof is a potential water source.
- Sun map—full sun, partial sun (3-6 hours) and shade. Track the shadows by going out and looking at sunrise, 9 am, noon, 3 pm and sunset. If you can add 6 am and 6 pm (assuming the sun is already or still above the horizon at the time), you'll get even more information. Map shadows during all seasons if possible especially in areas with long winters.
- Zone and sector map(s) (see the next section).
- Soil map—what kind of soil you have, where—even in a small garden, there may be a fertile corner where someone kept chickens! Take samples wherever you see a difference in what's growing there, or where you can see the ground is wetter or stickier, or where

you can see the soil actually looks different.
- And finally, the plan!

ZONES AND SECTORS

Zones

Zones are a way of thinking about our needs and labor; *sectors* help us think about limiting factors. We need to overlay both of these on our map to get the ideal design.

For zone planning, you're thinking about how often you use a particular area or element. For instance, if you eat salad and tomatoes nearly every day but harvest nuts once a year, you need your salad plants closest to the house and you can put the nut trees further away. Herbs need to be close to home; apple trees don't. Chickens need to be fed daily, so they need to be close to the house; your pond doesn't.

Zones run outwards from your home. Zone 0 is your home. This is a given for most of us.

Then, zone 1 is the most intensively used zone. It needs to be not just close to your house, but easy to access—so that narrow strip down the side probably isn't included. Zone 1 will probably include the kitchen garden, wormery, greenhouse, and rainwater tanks. This is where you'll do full mulching and maybe even use drip irrigation. In fact, this isn't a hundred percent permaculture; it's a human-made culture that would eventually be taken over by other plants if you didn't look after it.

If you only have a balcony or a tiny backyard, you'll only have zone 0 and zone 1.

Zone 2 is not quite as intensively cultivated. It will include perennial vegetables with a long growing season, maybe orchards, compost heaps, beehives, your chickens, vegetable beds that are sheet-mulched or green-mulched.

Zone 3 is the 'home farm,' with more extensive orchards and maybe large livestock (sheep, goats) and bigger trees. It probably doesn't get watered and has green mulch.

Zone 4 is partly managed but partly wild; it could include animal forage, timber production, wild foods. You manage it by grazing or by thinning out seedlings, but it's not 'garden' anymore.

Zone 5 is a natural ecosystem. This is usually further away from the house, but if you want to see wildlife, you might decide to allow a wedge of it further into your garden.

CREATING YOUR PERMACULTURE HEAVEN   23

AN EXAMPLE LAYOUT OF A BALCONY GARDEN USING PERMACULTURE PRINCIPLES

ZONE AREAS

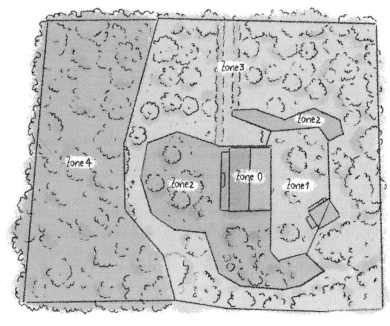

ZONE DIAGRAM

Sectors

Sectors are inputs—energies coming into the site, be that from wind, rain, water, fire, sun, or animals. They could also include local resources like a shop that throws cardboard boxes and pallets away. By mapping the sectors, you can see how you can close the gaps in the system or mitigate the problems.

- Is water flowing too fast through the site? Can we slow it down and plug the leaks?
- Is the wind too strong? Can we put up a windbreak—maybe a temporary one while the garden is getting established?
- Is the sun too hot in a particular place? How can you introduce shade?
- Block potential avenues of fire with pools, marsh, stone walls, or a firebreak.

As well as blocking energy, you can open up to incoming energy. Get more sunlight by clearing trees that are too tall; open up a view if your garden looks out towards the sea or the mountains.

You've done all this work so that your garden design is tailor-made for your particular circumstances and energy-efficient in terms of using what nature gives you to achieve the most with the least effort. Even little things count; if you're growing mulch plants, put them uphill from the beds where you're going to use them. Put tall trees at the top of a slope to stop cold air from rolling down it at night. And let your paths run along the contours; this will help avoid soil erosion.

The zone system

Zone	Purpose	Structures	Water sources	Animals	Plants	Techniques
Zone 1— This has the most intensive use and needs to have good access (zone 0 is the house)	Provide daily food, social space	Greenhouses, cold frames, path to house, shed, storage, worm bin, workshops	Rainwater tanks, water bores, wells, greywater, outdoor tap, indoor tap	Small animal pens (rabbits, guinea pigs for example)	Kitchen garden for vegetables with short growing season, small fruit trees, garden beds, lawn, flowers, herbs, small shrubs	Sheet mulching, propagation, drip irrigation, raised beds, bio-intensive beds
Zone 2— used quite intensively but not as much as zone 1. Larger and less frequently used elements	Food production, bird habitat	Compost bins, greenhouse, tool shed, wood storage	Ponds, well, large tanks, swales, greywater	Bee hives, chicken/poultry enclosures, rabbits, enclosures for larger animals that need to be regularly monitored and attended to, fish	Perennials with longer growing season, fruit trees, orchards, multifunctional plants, nuts	Sheet mulching, spot mulching around trees, drip irrigation
Zone 3— this area requires minimum maintenance and care once established. Farm zone.	Main crops for personal use or to sell, firewood and lumber	Feed storage for animals, shelter for animals	Dams for water storage, drinking water for animals, large ponds, swales, soil storage	Bee hives, livestock are kept and grazed, eg cows and sheep, semi-managed bird flocks. Pigs, horses, free range poultry	Orchards, main farming crops, large trees for animal forage	Green mulching, plantings are unpruned and not all have irrigation to water them. Cover crops
Zone 4— part wild/part managed	Collecting wild food, timber production, pasture for grazing, animal forage	Animal feeders	Ponds, swales, creeks	Grazers (deer, pigs, cattle)	Trees, pasture plants, trees for firewood and timber	Thinning (removing) seedlings to select the variety of trees to grow
Zone 5— unmanaged wild natural ecosystem such as bushland or forest. Free of human intervention. In urban areas it may be a creek or neglected land	Foraging, wilderness conservation, learning from the cycles of nature left untouched, meditation, connect with nature	None	Lakes, creeks, rivers	Native animals	Mushrooms, native plants	Unmanaged

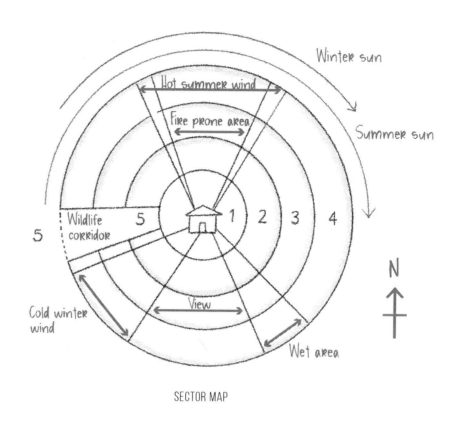

SECTOR MAP

Specific questions/observations to ask in the design phase

Questions

What services are provided by local government?

Are there any nearby plants that may affect our site now or in the future?

What is the location of structures on-site? House, garage, shed, walls?

What are your neighbors' activities? Will these affect your site? Will your activities affect them?

How big is your garden?

How far is it from your house to the furthest points?

Where are the access points? How easy is it to get materials, tools, etc., onto the site?

CREATING YOUR PERMACULTURE HEAVEN   27

Where are the power sources?

What is the wind direction and intensity; are there seasonal changes?

Do you have windbreaks?

Do you have damp areas? Or problems with water erosion?

What is the slope?

Are there any resources available in the community such as mulch materials, tools, machines, expertise?

Where are the water-catchment areas?

Where are the shady areas and the sunny areas?

Observations

Temperature highs and lows, dates for first and last frost

Rainfall amounts, seasonal variations, snow

Sunrise and sunset times over the year

Topography—the shape of the land

Geology—the underlying rock

Soil type, compaction, drainage, erosion

Areas where water pools, muddy or damp areas

Traffic outside, access for bringing in tools and plants

Water movement, flood zones, drainage

Views, privacy issues

Species on-site, health of the system and of plants/animals/insect life

Siting of any large trees or other sources of large shadow

Beneficial animals, pests

Zoning or planning restrictions, building regulations (if vertical gardening/roof gardening)

Leasehold or condominium rules if relevant (e.g., for balcony gardens)

Your resources—people, tools, time, money, local societies, plant exchange schemes, etc.

Check for underground cables and gas lines etc before digging.

MOVING ON—UPDATING YOUR MAP AND YOUR OBSERVATIONS

Keep a diary. It can be useful to have a standard format for items like rainfall, temperature, river level, sunlight hours, frost, so you can notice any patterns. Write down what you do in the garden and the results.

Take plenty of photos. If you use a standard viewpoint and time, this can give you a really strong feel for how your garden is growing.

It's worth talking to neighbors and friends who live nearby about their observations, particularly if you know people who have already started working in a sustainable way. Internet sources may give you average rainfall and other climatic statistics; local natural history societies or museums may have data on local flora and fauna. And you may find your site has a history. For example, there used to be a barn in my garden. It's gone, but where it stood, there's a lot of rock in the soil.

ENVISIONING—CONNECTING YOUR VISION WITH THE OBSERVATIONS

You need to have your own vision for your garden. Obviously, you're interested in permaculture and so you will build into it the basic permaculture principles; leaving the planet a better place, conserving and reusing water, using fewer fossil fuels, creating a habitat for many living species, producing enough food so that you can help others, organic farming with no pesticides or fertilizers, creating a space where you have to do less work, etc. But your vision might be different, depending on your circumstances and where you live. If you look at the visions for a few gardens, they will all be different, for instance:

You might come up with a vision statement that's a lot longer than that. Remember, in permaculture, we're working with the land as an equal partner, so ideally, your vision should be about your land as well as about your own needs.

Then you can look at specific wants. We'll cover those in more detail under planning, but for instance, you might want willow for basket making, you might want fruit trees, you might want fresh salad all year round, a wildlife area, room for chickens or livestock. You'll definitely want a compost heap, but what about a shed, or a play area for children, or a fire pit for socializing?

Remember, if you miss something out that you later realize you want, it's going to be much more difficult to find a place for it.

CREATING YOUR PERMACULTURE HEAVEN 29

Try placing these elements on the map. Look at the flow for your various activities (e.g., fetching eggs, harvesting salads). Grouping 'activities' together works well. The compost bin needs to be close enough to the house so that it's easy to fill, but it also needs to be close to where the compost will be used; get this wrong and you will have a lot of long wheelbarrow trips to make.

"I want a relaxed but productive garden where I can watch my chickens and my lazy old cat."

"Unless I do something smart with my water resources, nothing's going to grow on my plot – we're practically in the desert. Maybe permaculture can help."

"I have a lot of gardener friends who use loads of fertilizers and pesticides. I want to create a permaculture garden as an example to show them a light-touch approach can really work."

"I want my kids to grow up knowing the taste of home-grown food and understanding the natural world."

"I want fresher food in my diet, food that I know doesn't have all kinds of chemicals in it. Even though I only have a balcony, I'm going to try to use permaculture techniques to grow my own salads, chili peppers, and fruit."

"I want birds, butterflies, frogs, toads, hedgehogs, ladybugs – all kinds of living creatures filling the garden with color and sound."

"I don't have a garden, but I'm going to take on half my mother-in-law's garden and share the crop with her. She's having a tough time keeping the whole garden under control all on her own, so that'll help her and it will help me too."

"I just need some greenery in the concrete jungle!"

"Nothing at all grows in my back yard. I need to really get the soil into good condition, because right now it's just hard pan. And I want some shade."

PRINCIPLES OF PERMACULTURE

EXERCISES

You can analyze what you've done later, but right now it's just about getting all your feelings and thoughts down on paper in whatever form expresses them best.

Walking the contours—start at the end of your property and set one leg of an A-frame down and mark it with a post. Move the other leg of the A-frame until you find a level point. Mark it with a post. Swing the first leg round and when you find the next level point mark it and so on until you get to the end of your plot or hit an obstruction. The best thing to use is a spirit level to try and find the level points and to see if the ground is going up or down. Check that the bubble on the spirit level is right in the middle. If it is, you've found the contour line.

You could also mark out the 'map' you are creating using rope laid out behind you from one point to the next so that once you're finished you can look back and see the contour lines clearly.

Visioning—get a blank sheet of paper and some colored pencils or pens and put them in front of you on your desk or table. Sit back and close your eyes. Think about what your permaculture garden will look like, feel like, taste like, how the seasons will affect it, how you will feel walking or sitting in it. Let your thoughts wander but try to get a feeling for the key values. When you're ready, open your eyes, and using words, symbols, pictures—however you can best express your vision—put it down on that sheet of paper. If you don't like drawing, you can do the same thing by choosing images online to create a vision board. Or use photos cut out from magazines.

3

PERMACULTURE DESIGN—DESIGN PRINCIPLES

PLANNING

Planning is where we begin to work with detail. It's a bit different from the traditional method of garden planning, where the ground is treated as a blank canvas because permaculture principles say we need to think about the location, orientation, soil and use of the garden. Then we need to choose a design that fits the landscape.

If, for instance, you have a garden that slopes downwards towards a river, you might, depending on how steep the slope is, need to install terraces, or you might need to work out how to stop water from washing away the soil and creating gullies. But, on the other hand, if you wanted a big lawn… that's probably never going to happen in a permaculture garden, but it certainly won't in this one!

You also need to think about the cost. Buying plants, particularly trees, can be expensive. If your budget is limited, you may be better off creating a one-tree forest (guild—a group of plants, microbes, and sometimes animals, creating a pattern of mutual support. Often, a guild is centered around one species, typically a large tree or shrub, and is named after it) with an apple tree, some redcurrant and gooseberry bushes and supporting ecosystem, rather than trying to fill the whole garden. Alternatively, you could think about whether you can get free resources such as twigs and prunings (can the neighbors help? Do you know a tree surgeon?) to make a hugelkultur bed—something we will talk about later, but as an introduction, this is a no-dig raised bed made from logs, compost, leaves, grasses, etc.

Working with little 'chunks' like this is a typical permaculture move. It's not just useful for spreading your budget over a few years, but it also gives you a chance to see what works and where you might need to make changes.

INFRASTRUCTURE

Start with infrastructure. Think about what the garden needs to work—paths, watercourses, vegetable beds, a pond, a shed, or maybe a greenhouse or a polytunnel? Think, also, about how you're going to be moving around the garden; the compost heap can't be too far away from the kitchen door, for instance, or you're not going to use it. Some paths may need lighting installed if you're likely to need to use them in the dark. Herbs and salad vegetables need to be close to the kitchen; fruits and large vegetables can be further away.

Key features

Now you can think about the various things you want in the garden. Do you want a suntrap for catching some rays? A shaded area for hanging out on hot weekend afternoons?

Most permaculture gardens are set up for maximum productivity and don't include aesthetics as a top priority. But that's not to say it can't be a priority for you. If a profusion of flowers brings joy to your life, a permaculture garden can help provide that with less work (because you won't have to 'bed-in' annuals every year) and a display that will keep replacing one flower with another all year.

Plants

Now you need to think about picking individual plants. Think about the big plants first, as planting trees and larger shrubs will be more difficult once you've already got smaller plants growing. If you can, buy from a local nursery that knows your area and what grows well there.

Then think about the smaller shrubs. You can get a lot of fruit from bushes like redcurrant, blackcurrant, and cane fruits like raspberries, as well as nuts like hazel and vines like kiwi and grapes. As soon as you put vines in, you're starting to use the vertical dimension of your garden. Finally, you want to think about the annual plants and the ground cover plants. For instance, tomatoes, squashes, peas, beans,

Features you might want in your garden:

A play area

Trees (fruit trees, nuts)

Firewood (or wood for craftworking, willow for basket making, etc.)

Paths

Keyhole beds

Raised vegetable beds or hugelkutur beds

Herb spirals

Pools

Watercourses

Areas for birds or animals (paddock, chicken coop, wildlife area)

Orchards

A seating/family area

Solar heating

Greenhouse

Shed

Summerhouse

Barriers (particularly if you live in an area where deer are a problem)

An edible hedge

Fruit patches

Rain tanks

Beehives

Shading/sunny areas

Flower beds

A food forest

Forage planting for livestock

A compost heap

A meadow

and peppers are all annuals, so you'll have to start from scratch each year. But perennials should give you a good yield too and can form the backbone of your vegetable garden—Jerusalem artichokes/sunchokes, perennial kale and broccoli, asparagus, sweet potato, New Zealand spinach, and yams.

Many herbs, like rosemary, thyme, mint, sage, chives, and some types of onion and garlic, are perennial. Creeping thyme works as a good ground cover. And though technically many ground cover plants like clover are annuals, they will self-seed prolifically, so you don't need to get involved with replanting.

Remember, you're picking each plant not just for itself but for its relationships with the other plants. So, for instance, a way to make your orchard look prettier and stop grass growing up to your trees and stealing nutrients away from them is to plant a ring of bulbs around the tree. That could be onions or garlic—but it could also be daffodils and tulips, giving you a joyous show of color in spring. The flowers will also provide early pollen for beneficial insects.

When I'm planning, I like to make a table with three benefits for each plant, and unless I really, really love a particular plant, if it doesn't have three good benefits for the garden, it doesn't stay on the list. This benefits table will also help you when you come to designing in detail.

Wherever you can use a biological resource instead of a fossil-fuel-powered or labor-intensive resource, do so. If you have a big area, a couple of goats or sheep will graze it better than a lawnmower and without using all that gas (though watch out for goats with an appetite for your vegetables—keep them well fenced in!). Goats will also produce milk or meat; though you may like yours far too much to envisage the latter. Their milk is extremely nutritious and makes excellent cheese. You can also move a chicken enclosure around in a 'tractoring' process we will talk about later, which can be another efficient way of clearing land.

Fertilizers should be biological—manure, compost, nitrogen-fixing plants, green manures. In the same way, you'll control pests by biological means. Some plants, mainly the umbellifers (carrots, cumin, dill, angelica, caraway, for instance) and asters (asters, calendula, dandelions, daisies, sunflowers, and tansy), attract beneficial insects such as ladybugs, lacewings, and parasitic wasps, which will eat pests like aphids. Or you could encourage frogs and lizards. For slugs and snails, get ducks; they love snapping up the slimy things.

Some people are very keen only to use native plants. But for instance, in the UK, some plants—including fennel, walnuts, chestnuts, and mulberry—were introduced by the Romans. Is two thousand years enough to make them 'native'? A permaculture garden is one that uses plants that are well adapted for your climate and site, but it doesn't demand that they are natives. You may feel, though, that you'd prefer to use native plants rather than exotics—in which case, that's part of your vision.

Of course, if you use a greenhouse or polytunnel, you can grow plants that would typically be suited to a warmer climate, but you ought to think about heating your greenhouse without wasting energy. For instance, you might use solar heating, or site your chicken coop next to the greenhouse so your hens' body heat warms it up, or put the greenhouse by a pond that will reflect sunlight into it. If using a chicken coop, the greenhouse will heat it during the day, so the heat sharing actually goes both ways.

DESIGN PRINCIPLES

While you are designing, you need to think about several basic principles.

1—Relative location matters

Place elements according to their relationship—inputs and outputs. Plants with a beneficial relationship grow next to each other (we'll look at 'guilds' later in the book—they're a bit of a shortcut to placing the right plants together). A south-facing wall can be used for growing fruit that needs sun; the chicken coop could go near the fruit trees because the chickens will eat up any windfalls and thus prevent fruit flies from bringing disease.

Even a prevailing wind is an element that you can use. If you plant wormwood upwind of your vegetable patch, insect pests will smell it, and because it's a strong scent and they really loathe it, they won't catch the scent of all those succulent vegetables just waiting to be ravaged.

2—Each element performs many functions

Remember that principle we've already talked about? One of the slight differences in how permaculture gardeners think is that they think about functions rather than objects. So instead of 'fence,' 'wall,' or 'hedge,' we might think 'barrier,' and then look at the different functions that it can fulfill.

A dry-stone wall with gaps between the stones can create a climate for tiny ferns as well as hidey-holes for lizards and other small animals. A hedge can be an edible hedge producing fruit and nuts, and it can be a windbreak for smaller plants—it's actually a better windbreak than a wall because some of the wind goes through the hedge and this reduces the strength of the wind and any air turbulence.

Rather than thinking about 'mulch' or 'compost' or 'a wormery,' we can think about 'recycling organic matter.' Although for many people, a compost bin is practically the distinguishing characteristic of an 'organic' or 'ecological' garden, many permaculture gardens don't need one because the recycling is being done a different way by green manures.

Remember that a permaculture garden is a living system, so you need to address the critical aspects first—water, sunlight, energy. Make sure you have the basics for managing these aspects in place before you think about individual plantings.

With plants, do your analysis before you decide where to put it. First, look at its form. How tall does it grow? What shape is it? Some shrubs sucker up, while other bushes more often grow on a single stem; a fan-shaped bush will give more room underneath it for ground cover plants. Is it annual or perennial?

Look at the plant's tolerances and needs. How much light does it need? How much water? Does it need a particular type of soil? That can be pretty important if you are just starting with a garden with relatively poor soil; later on, after a few years of regenerating your soil, it won't be so important.

Then look at its different uses—to you, other plants or animals, and the soil. So a plant might produce edible fruit for you and provide nitrogen-fixing for the soil (like peas and beans do, for example); some reeds can purify greywater, as well as providing materials for basketry.

Some people like to call this analysis 'needs and yields.' Let's look at a slightly more complicated example in practice. Planting a grapevine on a trellis on the west of your vegetable bed will provide shade from the scorching afternoon sun; grapevines have space underneath, and so you can plant strawberries there. And in the autumn, the leaves will fall, and you can use them as mulch on the strawberries and the vegetables.

Of course, the strawberries are utterly useless unless you like eating them.

3—Each function is supported by many elements

Redundancy creates resilience. It's a basis of IT networks; always have two ways a link can be made, so if one doesn't work, the other will.

For instance, grow several different varieties of tomato, so if one fails, you still have a harvest. Farmers who grow monocultures are much more vulnerable than farmers who grow a variety of different crops for this reason. Mexican farmers bred thousands of distinct varieties of corn for different purposes or different environments; that huge biodiversity protected them against the loss of any single variety. By building more links between elements and having a greater number of elements, we secure resilience if any single link fails.

For instance, with water, unless you're off-grid, you have mains water, and you'll also have rainfall. But if the mains is cut off and it's not raining, you have to hope that you and your plants can make it through till the service is restored. Add a rainwater tank that captures rainfall from your roof, and you have a reserve; run it through a filter, and you can even drink it. Add a greywater-harvesting system that takes your bath water and laundry water, either direct to your trees (not to the vegetable bed) or through a reedbed filtration system, and you've got yet another reserve. Add a pool or two. Add swales if you're on a slope so that water is slowly absorbed into the ground instead of running downhill and away when it rains. You now shouldn't have any water problems.

Think about your aim of producing food all year round. Choose different varieties of fruit trees, some of which produce early fruit, others of which produce fruit later. If you have a late frost, your early bloomers might not bear fruit, but you still have the late-blossoming trees to give you a yield and provide food for birds. With vegetables, extend the season by using your window shelf or greenhouse to get springtime vegetables started and grow melons and tomatoes inside to pamper them through the first frosts. Again, you're giving yourself more resilience.

4—Establish energy cycling

This is right at the heart of permaculture's vision of a world in which we use the cycles of nature to work for us instead of trying to control them. Energy cycling tries to take energies flowing *through* the garden (in one end and out the other) and instead make them cycle *inside* the garden by capturing and storing incoming energy, like rain or sun, and recycling fallen leaves and cut plants as compost or mulch.

Think about how your garden will develop in the long term. This is something we're not very good at; TV programs give us the idea that you can create a garden in a couple of weeks by going out and buying lots of stuff, using a big earthmover and lots of decking and hard landscaping, and *voilà!* But we know that plants don't stay still—they grow. So do environments.

5—Adapt to succession

Forest succession is the natural progression from an open field, to ground cover, to the establishment of perennial plants, then shrubs and eventually trees. The first plants will stabilize and start fertilizing the soil; ones with deep tap roots bring up nutrients, others fix nitrogen, and thorny plants protect the others from being overgrazed. This opens the way for the next species, the shrubs; eventually, shrubs will take over. They provide an environment for trees to grow, till eventually, you have all those layers from tree canopy at the top to ground cover at the bottom.

This can take fifty years in nature. But we can accelerate it through wise choices, while natural progress will still work in our favor. By planting all the elements of the forest succession at once, we can create that multi-tiered structure—we don't have to wait.

6—Use the power of the edge

Edges are places where things happen. Along the water's edge, or where wooded areas turn into grassland, you get an 'ecotone' area in the middle which has extra-high biodiversity. Design your garden with plenty of edges by having lots of different communities that transition into one another.

Look at a photo of any river as it runs a winding course through the landscape. Copy this natural design for an edge: design ponds with inlets and peninsulas, curved paths instead of straight ones, keyhole beds. Alternate crops ('edge cropping') or use wavy, not straight, lines of veg.

7—Use patterning

Permaculture copies natural patterns, like the branches of trees, waves, and spirals. These are formed by growth or flow; they are dynamic, not static. We find waves in the ocean, but also in sand dunes; spirals can be as big as a hurricane or as small as a fingerprint. Many flower heads are patterned as two spirals going in opposite directions. Branching patterns make trees breathe, but they do the same for us; our lungs are brachiate. Sometimes, patterns are formed by shrinkage, like net patterns in cracked ground.

EXAMPLE OF A POND WITH EDGES

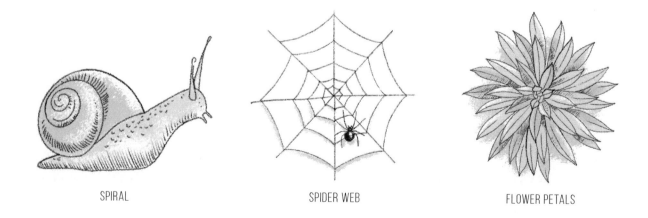

SPIRAL — SPIDER WEB — FLOWER PETALS

VARIOUS PATTERNS

We also find the Fibonacci series and the golden section in nature. Lots of natural patterns incorporate one or both of these patterns. The Fibonacci series adds each number to the last, so it goes 0+1 = 1, 1+1 = 2, 2+1 = 3, and then onwards to 5, 8, 13… The golden section or golden ratio is about 1.6, and each Fibonacci number is about 1.6 times the one before (e.g., 34 divided by 21 s 1.6; 5 divided by 3 is 1.67). Fern tendril spirals and branches both use this sequence as part of their geometry—spiralling out as shown in the picture below, for instance. Nautilus shells have the same basic geometry.

We find hexagons in tortoise scales, honeycombs, snowflakes; we find spheres in soap bubbles and dew and rings in lichen, mushroom 'fairy rings' and tree rings.

Is this just about permaculture aesthetics? If you like straight edges, can you still do permaculture? Well, yes, you can, but there are some real advantages to following natural patterns. For instance, many types of patterning can help you plan to take the best advantage of the space you have, such as branching, which lets you organize a garden around branching paths, so you have fewer paths and more garden. That can save labor, too, by connecting the various focuses of your outer-zone edible forest to each other, so you don't have to come all the way back to the house each time you want to move from one to another.

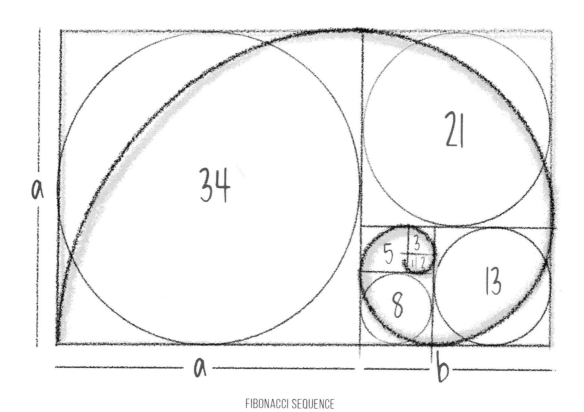

FIBONACCI SEQUENCE

If you're short on ideas, Geoff Lawton has loads of great videos on YouTube. You might also want to look up the 'original' gurus of permaculture, Bill Mollison and David Holmgren.

Let's mention a couple of specific styles of patterning.

Keyhole beds

When you stand on soil, you compact it, squeezing the oxygen out and undoing all the good work the microbes and earthworms are doing. So many gardeners use raised vegetable beds, where they can reach all the vegetables without having to walk on the bed. But they're usually rectangular, with a limited amount of 'edge.'

Some permaculture gardens, instead, create a circular bed with a small path into the middle—a 'keyhole bed.' You could use a horseshoe shape, too. If you raise them slightly in the center, you can plant the taller vegetables in the center and those needing more sunlight on the edges; or even let squashes and cucurbits drop down the wall.

They're simple to lay out with just a bit of string tied to a stick. You can use wicker, stone, or chicken wire for the sides; you can even make a 'mandala' garden with a whole circle of keyhole beds around the center. Some gardeners in arid climates also add a cylindrical bin that can be used for both composting and watering at the center.

Design principles

Relative location

Each element performs many functions

Each function is supported by many elements

Energy cycling

Succession

The power of the edge

Patterning

PLANTING IN ROWS IN A TYPICAL GARDEN

KEYHOLE GARDEN

KEYHOLE GARDEN IN ZONE 1

Herb spirals

These are another way of creating edges and microclimates and consist of a conical mound of earth, with rocks or bricks making a spiral path around it. Each herb can be planted according to its needs; Mediterranean herbs like thyme, oregano, rosemary go on top where it's dryer; herbs that prefer shade and moisture, like mint, can go on the south-east, at the bottom. The heap even insulates itself to keep the soil at a steady temperature. And it's a very attractive method of planting. People often plan these in zone 1 near the house, for easy access.

DEVELOPMENT/DESIGN

Now you're ready to get stuck in and actually design your garden, creating a detailed plan for laying it out.

How you design your garden will depend on what makes the most sense for your

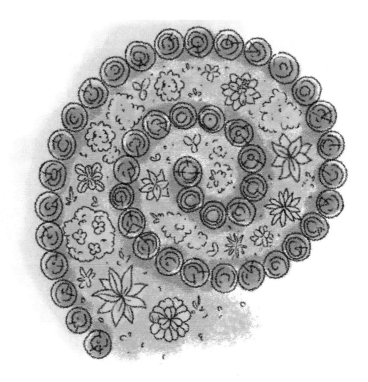

HERB SPIRAL IN YOUR GARDEN

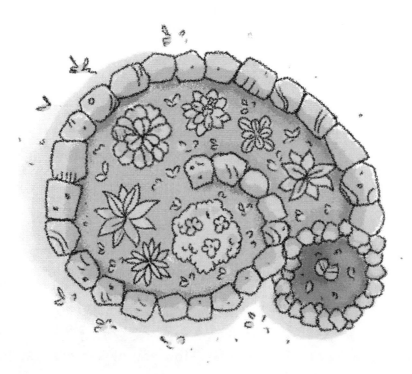

HERB SPIRAL IN YOUR GARDEN

particular piece of land and the climate and ecosystems that already exist there. If you're starting in an arid area with no existing planting, you're going to have to think about water resources and shade first, and you'll probably find the soil needs a lot of help. You might want to see if you can create shaded microclimates, and you might also want to concentrate on getting shrubs and ground cover started rather than planting trees. Trees are expensive to lose, and they'll be much happier if they have shrubs to protect them while they're young.

On the other hand, if you've inherited an orchard with the trees marching in straight lines (not very permaculture-friendly, but that's what you've got) and the soil has been depleted over the years by growing the trees in bare soil or grass, then you may want to think about other issues. Should you treat each tree as the center of its own mini food forest or alley-plant shrubs and vegetables between the rows of trees?

And evidently, if you're planning a farm and there isn't a house on-site, you'll need one (or a camper van or static mobile for the moment). If you want animals, you'll need fencing, or you will be sharing more of your harvest than you want. And all of this needs to be done before you start planting.

Also, think about cost. Some resources come free with the land, and some can be begged or obtained for very little; other things like plants will cost (though if you are short on cash, look through Freecycle and community noticeboards). Remember I told you about the barn that used to be on my land? I hoarded all the stones I dug up, and I had enough for two herb spirals and a bit of wall.

You might want to use a bubble diagram—draw the features you want (vegetable bed, orchard, greenhouse) as bubbles on the map, giving a very rough idea. Or you could use post-it notes and move them around as you need. This is just a concept design, asking what *kind* of stuff goes where. We're not going to think about individual plants and trees so much yet, just where you need a hedge or a line of trees, where a pond might be useful, or a sun trap.

When you're looking at the map, connect things up; don't get stuck with a single idea. Think about how each element connects to the others. Think about flow, how you can move through the garden; think about the sectors and slope and watercourses and how those affect the design. Check your ideas against the zones and sectors you've already drawn. You may find that you're moving things around quite a bit until you find the right locations.

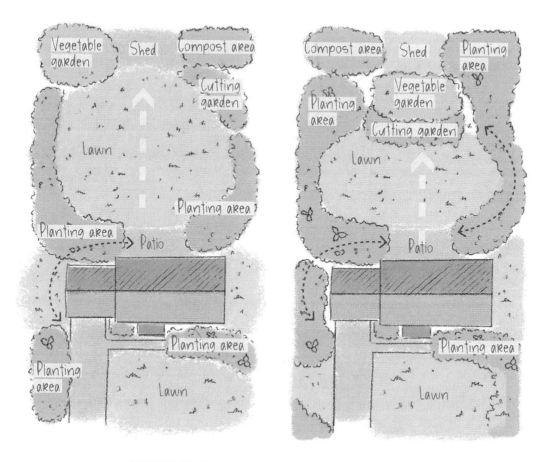

VARIATIONS OF A BUBBLE DIAGRAM USED TO DESIGN YOUR GARDEN

Add your swales if water runs down a slope—they will not only stop soil erosion but also hold water long term, and they're good places to put bushes and plants that need moisture. Set up rain-capture from the roof or a greywater reedbed for purification. Set up shade and sun areas and see if you have possible areas for microclimates for different plants.

By the way, take note of where any power lines are. Don't plant trees near them that are going to be big enough to touch the wires when fully grown. You might want to check with the water utility where any underground pipe has been laid too. Likewise, find out if there are any buried electric cables anywhere on the land.

Once you've got your map done, it's a great reference document. It will help you for years to come.

The next dimension of development—time

Now you need to plan the installation, that is, what jobs to do and what to do first. This may depend partly on your budget; if you can only get started with one little cluster of plants or a single vegetable bed, start with the best place and easiest setup to get started, or with the vegetables that you'll eat most often (if feeding the family is a priority, annuals, tomatoes, and fruit bushes will all start to yield long before fruit trees).

Make notes on how long plants will take to grow and to transform the soil. If you're starting with a very bare, arid site, it could take several years of working on water reclamation and soil regeneration before larger plants can be introduced. If your garden is an overgrown, abandoned patch, you'll have a lot of chopping back of plants to do before you can start planting. That might make you choose to use the twigs and branches for a hugelkultur bed in the first few years (this will be discussed in a later chapter), recycling your waste and fertilizing the soil.

A lot of what you'll be doing is probably cross-checking functions, needs, and yields—which plants need sun, shade, moisture, which are nitrogen fixers, and so on. Most gardeners start in zone 1 and work outwards and unless there are special circumstances, plant the critical larger plants early, as they take longer to grow.

For large permaculture, you'll want to take things in this order:

1. Infrastructure and tools (greenhouse, access).
2. Erosion control, earthworks, water, fencing, or barriers.
3. Pioneer species ('weeds') to stabilize soil, fix nitrogen, accumulate nutrients and use as green manure or mulch.
4. Perennials—trees, shrubs, windbreaks.
5. Annuals.

Don't spread yourself too thinly. Try to divide your plan into separate work packages and get each piece complete before going on to the next. For instance, if your first job is to get annuals like salad plants and herbs growing close to the house, get that done first. At least you can have a chive omelet and mint tea while you're working on the next bit of the garden! The outer zones come later.

Build up over time. Use the seasons and the principles of succession to help you.

Whatever you do, DON'T bare the soil! If there are loads of weeds, use them as a green mulch; just chop them down and leave them there. You can sow green manures through the weeds or dig holes to plant through. That will help recondition the soil. Then it's time to plant the perennials.

Or you can lay down sheet mulches (discussed in more detail in a later chapter) or make a hugelkultur. This creates fertility fast, and you can plant greenhouse-grown plants such as courgette, tomato, or squash in little pockets in the mulch. But sheet mulch needs some time to rot down till you can plant—it's best to mulch in fall and plant in spring.

DO YOU WANT TO DO A DESIGN COURSE?

Some people learn well from books. Others prefer to take a hands-on approach to learning and go on a course, either because they can ask questions and get some help with their own circumstances, or because they have a more kinaesthetic or aural learning style, or simply because it gives them more confidence.

There are plenty of permaculture courses available, but you need to choose one that's in line with your circumstances and approach. Look at the teacher's biography. If you're a young, urban professional wanting an urban rooftop and backyard garden you can hang out in and get some food from, too, don't pick a course with someone whose expertise is in big agroforestry projects. If you're struggling with an almost desert environment, don't pick someone who farms in a northern mountain state.

Make sure when you read the instructor's writing or watch their videos that it resonates with you. If you can't stand new-agey stuff and the instructor keeps talking about nature's great design and the spirituality of indigenous peoples, head for the one who talks about tick lists, order quantities, and costings. You may like a teacher who is funny, sparky, who teaches by telling stories, or you may want a more structured environment.

Be specific about your goals. What are the things you don't quite understand or lack confidence in putting together? Many people start with the Permaculture Design Certificate, and that's a seventy two-hour slog that will let you set up as a professional, but maybe all you really want is a specialist weekend on plant guilds or water management.

Look for like minds. African-Americans and particularly female POC are often in a minority of one on permaculture courses; check out the Black Permaculture Network and Pandora Thomas at the Earthseed

Permaculture Centre for a view of permaculture that takes human networking as well as natural systems into account.

And finally, I would really recommend you get a hands-on, real-world course, even if it means a bit of traveling and expense. A Zoom meeting or online video can't replace being able to smell what a compost heap *ought* to smell like, crumble earth in your hands to find out its quality, or actually walk around and experience a successful permaculture garden in all its beauty and fertility.

EXERCISES

You can analyze what you've done later, but right now, it's just about getting all your feelings and thoughts down on paper in whatever form expresses them best.

Patterning—look at pictures of different natural patterns in books or on the internet. Are there particular patterns that resonate with you? Think about how you could incorporate them into your garden.

A miniature exercise—setting up balcony permaculture might sound easy. But go through all the stages of the design. For instance, sector analysis—how much sun do you get? Where does the wind come from? If your permaculture's going to be a bigger one, this is a great way to think about the process before you carry it out for real on your own site. If you actually have a balcony, congratulations; once you've done this exercise, you're ready to get started!

4

CATCHING, CONSERVING AND USING WATER

We're used to thinking of water as coming in through the tap and going out through the drain. That's an incredibly wasteful way to use a scarce resource.

It's true that seventy-one percent of the planet's surface is water. You'd think that was more than enough. But ninety-seven percent of it's in the ocean. That leaves just three percent of the planet's water, and the great majority of that is locked up in glaciers or permafrost, leaving just a third of a percent of the total water supply for use on land. So rivers, lakes, warm water vapor in the air waiting to be cooled into rain, water held in the soil or underground, and water as part of living things (like us—we're about sixty percent water) add up to about a third of one percent of the total. In fact, even if we looked at the more general class of freshwater as opposed to saline or brackish water, that's just 2.5 percent of the total water available on the planet. So freshwater that's available for plants to use is a surprisingly scarce resource.

And the human population, which is a big water user, is growing steadily, while water resources aren't. So we really need to strive to keep our water use down by carefully harvesting, storing and recycling water. Water management is a big part of permaculture, in line with the principle of caring for the planet.

Normal gardening is wasteful with water. It leaves the soil bare, so water evaporates; it uses sprays, sprinklers, electric pumps and hose pipes. It's power-hungry and takes its water from the tap. Yet, even with irrigation, gaps in sprinkler coverage, or low water pressure, can lead to under-watered areas. Other areas are over-watered, which can kill plants if their roots are sitting in water.

The permaculture gardener, on the other hand, thinks about ways to preserve water. Mulching stops water from evaporating from the soil. Swales, terraces and berms along the contour lines of a slope stop water running downhill and let it soak into the soil instead (we will discuss these in more detail). Plants along the contour will be able to pull from those reserves if there's a drought.

Plants like enset (wild banana) with deep roots are planted to reach down further for water. Some gardeners harvest rooftop rainwater and greywater (discussed later in the chapter) for use in the garden. Pools and ponds hold reserves of water.

In dry climates, stones placed around trees reflect heat up to the foliage but create a cool, damp micro-climate underneath and unglazed pots are placed in the ground, then filled with water to release it slowly to the roots. In Mexico, chinampas—floating islands made of lake mud and rotting vegetation—allow a kind of natural aquaponics that's incredibly fertile.

When you're planning a permaculture garden, you need to look at your own micro-climate. It can be very site-specific; just a few miles can make a difference. Start with the baseline for the nearest city and set up a spreadsheet with a row for each month and columns showing rain, notable weather such as storms, and average temperatures. But then, start taking your own records, using a rain gauge and a thermometer. Take the temperature at the same time every day, or even better, several times a day. You may find your figures are quite a bit different from the baseline.

Even in quite a wet climate, you can never rely on rain just when you need it, so you need to store water. For resilience, don't rely on any single source of water; try to have reserves in different places from different sources.

And don't grow plants that aren't suited to your climate (or micro-climate). Put plants where they'll get the amount of water they need—no more, no less.

AN EXAMPLE OF PLACING STONES AROUND THE BASE OF THE TREE TO CREATE A MICROCLIMATE IN DRY CLIMATES

STORING WATER IN THE GROUND

When soil is left bare, water easily evaporates. So plant your garden densely, so there's no bare soil between plants and the soil is protected from the sun and from weeds taking over.

If a bed isn't being used, then you can either plant a cover crop, like alfalfa and use it as green manure or compost later, or you can mulch. Mulch is a layer of material applied to the surface of soil to conserve water and improve the fertility and health of the soil. The deeper the mulch, the better. Old carpet (if it's wool or cotton), cardboard boxes, newspaper, tree and hedge trimmings, or leaf mold can all be used, or even rocks (more mulch materials discussed later in the chapter). The mulch stops the water from evaporating and also cools the soil, which further helps retain moisture.

The benefit of an organic mulch is that it will, eventually, rot down into organic matter. The more organic matter the soil contains, the more moisture it can retain. While manure and mulch are great *materials,* they won't do your soil any good until they have rotted down into humus, that is, organic *matter*. And they shrink—ten parts of organic material becomes one part of organic matter—so you can see how thick a layer of mulch you're going to need to make a difference.

For every one percent of organic matter content, soil can hold 1.5 quarts of water per cubic foot of soil, down to twelve inches deep. Organically rich soil is like a sponge and can hold up to ninety percent of its weight in water. It also absorbs water quickly, so sudden rainstorms are less likely to leave water pooled on top of the soil. If you live in an area with highly changeable weather, organically rich soil is great because it acts as a water buffer, absorbing water when the weather is wet and retaining it when you have a dry spell.

Organic matter doesn't just conserve water. It releases nitrogen, phosphorus and sulfur, (mainly in spring and summer) and boosts soil life—microbes, bacteria, worms—which help to aerate the soil. Plantlife sequesters carbon, removing carbon dioxide from the atmosphere and storing organic carbon in the ground. Forests can sequester more carbon than grasslands before becoming saturated, according to a study from Colorado State University. That's a good reason why in a permaculture garden, following natural succession towards the full-grown forest helps the climate as well as your garden's own fertility. Wherever you have the full range of plants from trees all the way down to annual plants, the soil will benefit from fallen leaves, different types and depths of root system, and the trees will absorb carbon.

CONTOURING TO CATCH WATER

Water can erode soil and rock—look at the Grand Canyon! If you have a garden or farm on a slope, contouring is one way to reduce water erosion and stop your whole garden from getting worn into gullies. Contouring aims to make the water collect in, or run slowly along, the contours instead of making a beeline for the bottom of the hill. Contours are lines of equal height that follow the natural curve of the land and contouring simply involves planting along these contours. This looks completely different from the usual modern farm, with everything in straight rows.

'Keylines' enhance the contours without using terracing, for instance, by keyline plowing. Instead of turning the soil over, this plow slices narrow cuts into the land along the contour; the slots practically pull water into the soil. The keylines will be enough to absorb most of the water most of the time, so you won't even see it, or you can collect the keyline water into pools at the end. You choose your keyline to maximize water-collection potential, opting for the longest contour you can get.

You can alternate annual crops, for instance, by rotating legumes with nitrogen-hungry plants like squash or corn. A University of Nebraska study showed that where contouring was the only change farmers made to their practice, they achieved a five to ten percent increase in crop yields—some of them up to fifty percent.

AN EXAMPLE OF KEYLINE DESIGN

BASINS FOR ARID ENVIRONMENTS

Catching water in a dry climate is a special art. For instance, trees are planted in a basin, which is then filled with mulch. The shape of the basin will collect rainfall and feed it towards the tree, while the mulch will keep the moisture in the soil.

In desert areas, catchment areas are created with soil ridges or rock walls to direct the rainfall into a corner, where there is an infiltration pit to accept the water and where the plants are grown. The soil needs to be at least four feet deep to make these systems work; the catchment area can be between ten and one hundred square meters. In Israel, V-shaped 'negarim' are used to grow fruit trees, but you can work with half-moons, lozenges, or contour bunds, too. The latter are very similar to swales but with more of an emphasis on the berm.

DESERT CATCHMENT AREA

SWALES

Modern cities see rainwater as a nuisance to be gotten rid of as quickly as possible. But as permaculture gardeners, we want to keep water in the garden as long as we can and make it work as hard as it can. Swales are a way of slowing it right down and pacifying its destructive tendencies.

A swale is a ditch laid out across the slope, along the contour, or just slightly off-contour—the latter will move excess water very gently down to one end of the swale. Pile the soil on the downhill side as a berm, a long mound (so the ditch plus the berm look like an 'S'-shaped curve). You can fill the swale with a little gravel topped up by mulch, so the swale doesn't show, but it will still work. The water will collect in the absorbent mulch and then filter through to the berm and the slope below. Silt, leaves, organic matter and nutrients will also be caught in the swale.

When you plant the berm, you are both protecting the soil (through the plants' roots) and helping to keep the plants well-watered. The berm has twice the normal depth of topsoil because it has both the topsoil that was already there and the excavated spoil from the swale on top. This makes it very fertile. The increased water retention will also make the ground more fertile and moist for several yards below the swale, in what is known as a 'water plume.' This water plume will grow from year to year.

Swales should be from one to three feet deep and two to six feet across, with from three to eighteen yards between swales, but there are no absolutes. In sandy soils, swales can be shallower; in clay-rich soils, deeper. Just be sure you don't make the swales too big; if you make too big a holding capacity, the ground can become boggy and the sides can even blow out.

In a heavy precipitation climate, build your swales closer together. Permaculture expert Toby Hemenway said if you get forty to fifty inches of rain a year—most of the UK would qualify, for instance—you'll need swales about eighteen feet apart. In drier climates, they can be further apart.

The steeper the slope, the closer together swales need to be (one guide is that each swale should be at the same level as the top of the trees on the berm of the swale below). If you have a plot with some steep and some gentle slopes, plan your swales to use the gentle slopes—particularly if you have a hill that rises steeply at one point, use swales below or above the steep part but not on it.

The maths for holding capacity is easy:

width x depth x length

So a swale thirty feet long, one foot deep and four feet across gives you:

30x1x4 = 120 cubic feet, or 898 gallons

That's the maximum it can hold before the water starts penetrating into the soil—say, in a serious rainstorm. The city of Naples uses very shallow swales slightly differently to take road water runoff and reckons a 6-inch-deep, 8-foot-across swale can hold 15 gallons of water for every foot of length.

Try to find the longest contour curving around your plot. That will get the maximum water absorption, so it's a good place to put your first swale and you can work out the placing of others from there.

Don't put a swale just uphill from your house; make sure it's ten feet or even twenty feet away. Otherwise, you'll be flooding your basement.

For full-scale farm berms, trees are essential to stabilize the berm and to shade the swale. You might also plant trees on the uphill side, particularly nitrogen fixers. Their roots will form conduits in the earth and enable the water to spread even further. In a small garden, this is less important, but you'll still need bushes.

Over the very long term, the swales will fill up with organic matter, but by the time your trees are fully grown, the soil will be fully hydrated and the swale will hardly be needed.

For a small garden, you can mulch across the entire swale and even grow fruit in it—put some gravel or big logs in the bottom of the swale to soak up the water and it will be there under the apparently level surface. Swales aren't always flowing with water sometimes, depending on your climate; they just spread out storm-waters, but are usually dry, so you can use them as access paths most of the time. You could gravel the bottom if you want to do this, so they don't get too muddy.

Putting well-rotted manure or compost in the bottom of the swale is a great idea, as the water will carry the nutrients in that compost all the way through the ground, up to several meters below the berm.

Good plants for the berm

Rosa rugosa
Elder
Geraniums
Comfrey
Fruit trees
Redcurrants
Strawberries
Rhubarb
Squashes
Sweet potato
Custard apple
Banana
Ginger
Pineapple

Bad plants for the berm

Lavender
Rosemary
Santolina (cotton lavender)
Salvia
Agave
Yucca
Aloe

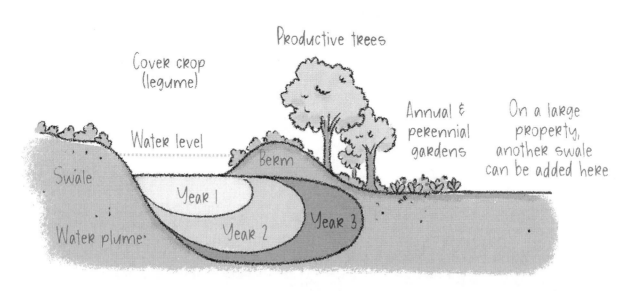

EXAMPLE OF A SWALE

Making a swale

Look at your map first and have a guess where you'd like the swale to go—along the longest contour you can find. You may need more than one, depending on the size of your plot. Remember, you don't want to build swales on too steep a gradient (more than fifteen percent).

Think about your average rainfall per year and how it comes—steady or in sudden heavy showers. If it's the latter, you may need to build larger swales further apart to hold the high volume of water and let it soak in gradually. Use the maths we looked at earlier to make sure your swale is big enough.

Make an A-frame. I like mine to be about the height of my hips for ease of use, so that means I need two pieces of wood a bit longer than that. I drill through both pieces at one end and connect them together loosely with a nut and bolt so they can swing out to make the sides of the 'A.' Then I simply open them up and attach a third piece of wood about halfway down, making the crossbar of the 'A.'

There are two ways to use it. Traditionally, a weighted string hanging from the top would be used to mark the exact middle of the centerpiece. When the plumb-line matches the line you've marked, the frame is level. Or you could just tape, tie or glue a spirit level to the crossbar.

Start out where you want to begin your swale and mark the beginning with a peg or a water-based paint. Using your A-frame, find the next point on the contour, where the A-frame is level (see above for an idea of distances they need to be apart). Mark that point, swing your A-frame round, then find the next point… and so on.

Once you've marked the swale, walk all the way along it. Does it *feel* right?

You'll need to do an infiltration test to make sure the ground drains well. Dig an eight-inch-diameter hole eight inches deep and fill it with water. Let it empty, then refill, start a timer and measure every thirty minutes how much water has gone. You need the soil to absorb at least a quarter-inch of water an hour, ideally.

Walk the length of the swale. Then use your fork to loosen and aerate the slope just below the swale. Find a crate, wheelbarrow, or other object from which you can gauge the width of the swale. Then dig into the side of the hill vertically on the uphill side, scooping the soil back onto the berm. You want the swale about three times wider than it is deep. Flatten the bottom with a rake and check it's flat with your A-frame.

Finally, even out the back-cut so that the slope into the swale is gentle and make overflow points—level edges where stormwater can overflow. You might want to reinforce these with rock or with willow or hazel sticks to prevent the spillway from being eroded.

It's hard work, so if you have friends who're into gardening, get a swale-digging party together. But remember, you only have to do the work once! Alternatively, get a professional in to help if you think it may be too challenging to do yourself.

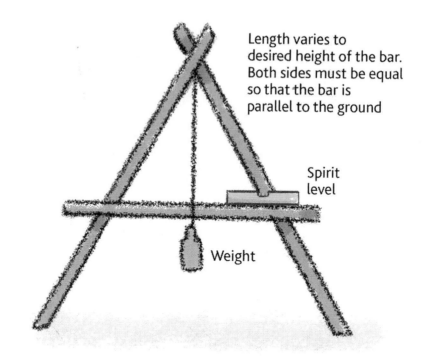

HOW TO BUILD AN A-FRAME

HOW TO USE AN A-FRAME

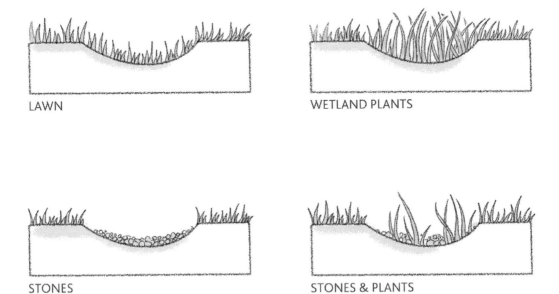

SWALE MATERIALS

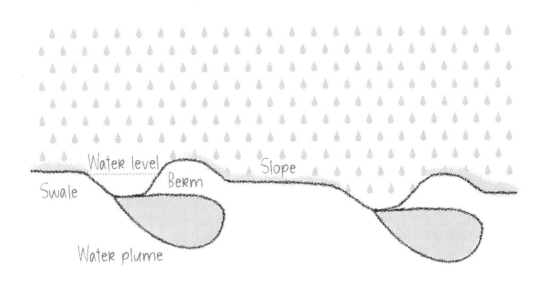

EXAMPLE OF A SWALE DESIGN

EXAMPLE OF A SWALE DESIGN

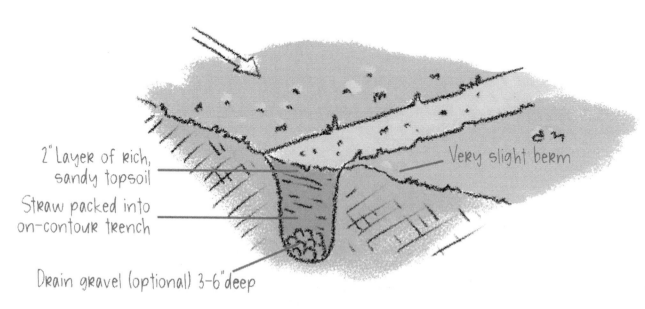

- 2" Layer of rich, sandy topsoil
- Straw packed into on-contour trench
- Very slight berm
- Drain gravel (optional) 3-6" deep

CLOSE UP OF A SWALE
TAKEN FROM 'GAIA'S GARDEN' BY TOBY HEMENWAY

Fish scale swales

On a complex slope with a lot of little hollows, fish scale swales create mini-contours. Mark the swales out first so you can see the pattern. Each swale forms a slight crescent or 'fish scale,' and your aim is to make them fairly regular.

Mini backyard swales

If you have a greywater system feeding a pond or reed-bed, your swales can then take the overflow water in a loop around the garden. Mulch them heavily so that the mulch will 'suck' water along.

FISH SCALE SWALES

You could also make a cascade of miniature swales. Fill them up with old cardboard and newspapers, then put an attractive layer of bark mulch or coconut fiber mulch that looks nicer. You can use these swales as paths.

RAIN GARDENS

Rain gardens are water-sinking depressions for dealing with torrential runoff in arid climates. They're designed to drain quickly and efficiently. You can start small and link them in tiers if you need to, with an overflow from one to the next.

As with swales, don't build them within ten feet of your house.

First, dig out a big scoop of earth, let's say eight by twenty feet long (you can rent a mini excavator at $150-200 a day if you need to, but of course, that will add to carbon emissions. Getting help from friends is more environmentally friendly and can also be much more fun). Dig it about two feet deep.

Then create a fast-draining soil mix by mixing sand and compost into the soil you dug out. You can also use a specially 'engineered' mix from a local supplier. Incorporating more sand is good for dealing with torrential storms if you grow desert plants like agave; on the other hand, a less sandy mix (forty-five percent sand, thirty percent compost, twenty-five percent soil) will support lusher plants, or fruit bushes such as beach plum, lowbush blueberry, or juneberry.

Fill the scoop with this mix up to about six inches below the surface. The gap is there so the water can pool if it doesn't drain immediately. You can disguise the gap between the levels by building a low wall around the rain garden.

In a rainy, hot, tropical climate, the *banana circle* is a variation on the rain garden. A pit is dug and filled with organic matter, then heavily mulched. A raised berm around the circle is planted with bananas and a mix of cassava, sweet potato, lemongrass and taro.

WHEN THERE'S NOT ENOUGH WATER

Soil can dry out rapidly, so keep it covered. In Texas' High Plains, many farmers now leave crop residue from their previous crop on the soil, either reducing or completely abandoning plowing to minimize evaporation. If they don't, high temperatures, together with the crop's own needs, can mean irrigation requirements go up to the equivalent of half an inch of rain every day.

Choose xeric plants that thrive on very little water. Plants native to your climate zone will do best; for US arid climates, prairie grasses and asters, or purple prairie clover (like orange butterfly weed, great for pollinators) are good species. If plants grow wild in your area, they'll also do well with the local soil.

Non-native plants from similar climates can also be used. For instance, lavender, rosemary and thyme are Mediterranean plants used to a dry climate and well-drained soil.

Building a resilient system of food production may not be possible with only native species unless you adopt a drastically simplified diet. You may also find that climate change has made plants native to your area no longer able to thrive in it, so you'll need to look further afield.

Some areas of your garden may have moister micro-climates. If there's a place where the lawn stays green when the rest is brown, there's water there. Put your thirstier plants near downspouts, in swales, or somewhere they can get runoff from sidewalks or paths (like curb cutouts). And put thirsty plants in zone 1 so that if they need watering, it's easy for you to do it.

But don't give up before you've got all the water you can out of the site. Permaculturist Geoff Lawton managed to make a desert in Jordan into a lush oasis!

PLANT DENSELY

Planting densely and using layers so that plants shade each other keeps roots cool and reduces water loss. A study from Maejo University in Thailand showed that the shadow area of perennial and ground cover plants could reduce the temperature by two to five degrees centigrade.

That's why a permaculture vegetable bed will use ground cover plants to avoid bare soil. Creeping thyme, oregano and clovers like a bit of sun; periwinkle, dewberry, wild strawberries, lungwort, sorrel and ajuga will grow in shade.

Beans, beets, broccoli, cilantro, leeks and onions all grow well in partial sun; peas will form less foliage when they're in the sun, so they're able to put more energy into the pea-pods. For partial shade, plant chard, radishes, and leafy greens like lettuce, spinach, cress and kale.

But try to put tomatoes in the shade and they just won't thrive. They need full sun, so train them up a trellis or make them the tallest plant in the vegetable bed.

MULCHING

Mulching is simply covering the soil up with a *(usually)* biodegradable layer of organic material. Non-biodegradable mulches like a layer of stones in dry climates will help to conserve moisture, but you don't get the added benefit of the mulch slowly rotting down and being incorporated in the soil, improving your soil structure.

If you think about a natural forest floor, it's usually covered up with all kinds of stuff—old pine needles, pine cones, leaves, fallen sticks and twigs, as well as herbs and shrubs and their fallen leaves. It's never bare. So in the permaculture garden, particularly in a food forest, you cover up the ground in the same way. But because this isn't a natural forest and you might be starting from scratch, you might use different materials; you could use bark chips, newspaper, compost, straw, green manure, or old corrugated cardboard boxes.

Remember that the reason permaculture copies what happens in nature is because it's a system, and it's a system that works. So it doesn't matter that you might use different materials. What matters is that you're covering the soil with a barrier layer of organic material. This minimizes evaporation and erosion and also stops unwanted weed seeds from germinating. It regulates the soil temperature so it doesn't get so hot by day or too cold by night. And depending on the kind of mulch you use, you can increase biological activity within the soil and add nutrients as the organic matter rots down.

This will save you time weeding, watering, fertilizing and pest-controlling. As usual with permaculture, you have to invest time in the beginning, but the investment pays great returns, saving you all the back-breaking gardening work a conventional gardener will still be doing in ten years' time, while all you mainly have to do is pick the fruit and veg.

Mulch also increases soil moisture and the length of time that moisture will be stored, so gardens can go longer between waterings (whether artificial or by rain).

But there are a few things you need to think about. First, in a moist climate, mulch can be a nice hiding place for slugs—don't use too much of it!

In the spring, a heavy mulch can prevent the sun from warming the earth as quickly as where the soil is not covered. So if you want to get your crops off to a good start, rake off the mulch and replace it once the soil is warm. Alternatively, lightly work the last year's mulch into the soil after you've harvested and then add a new layer for the new year once the soil has warmed up.

In dryer climates, you'll need to mulch more thickly to retain more moisture, but in wetter/cooler climates, the mulch needs to be thinner so that it still helps protect the soil from erosion but still allows excess moisture to evaporate to reduce fungal issues.

Methods to retain water in the soil and their benefits	
Method	Benefits
Organically rich soil	Water less likely to evaporate
	Adds nutrients for fertility
	Stores nutrients
	Sequesters carbon
Minimise bare soil	Shades soil
	Less water evaporates
	Smothers weeds
	Keeps soil cool
Mulching	Maintains organic matter
	Adds fertility
	Protects soil from the sun
	Cools the soil
	Smothers weeds
	Deters some pests
	Warms soil in spring
	Protects roots from extreme hot/cold
	Softens impact of rain on ground
Soil contouring	Reduces soil erosion
	Reduces flooding
	Maximum water collection potential
	Directs water where needed
Swales	Attracts rainwater to soak into swale and fill the downslope
	Absorbs surplus rainfall and prevents run off
Rain garden/banana circle	Deal with torrential rain
	Allow absorption and drainage

TOO MUCH WATER

All the techniques that help when there's too little water can also help when you have too much. For instance, humus-rich soil will absorb water much better than compacted, clay-rich soil.

However, when organic matter is totally waterlogged, it doesn't break down well; if there isn't enough oxygen around, the little microbes that do the work can't survive. So your first point of call, in the worst situations, might need to be bringing in some new topsoil.

Clay is made of little platelets that stick together (you can see how this works if you get a couple of pieces of glass and wet one of them; the other will stick tight to it—it'll slide, but it won't separate). That's what clay soil is like. There's no room in it for water to be absorbed.

Once the soil incorporates humus, it clumps into aggregates which are larger and more irregular and have larger spaces between them for the water to pass. That enables the soil to hold a lot more water.

Using plants with deep taproots, like dandelions, comfrey and carrots, parsnips, or potatoes, can help break the soil up. In general, the more roots you have, the better, so even sowing a whole backyard full of weeds is better than leaving things the way they are.

In a very rainy climate, don't build swales on the contours, but at about a two percent grade. That will direct water where you want it, without it running fast enough to scour out a gully. Remember, though, that the swale should have a level spillway somewhere along it in case it becomes over full. This will enable it to release water safely without damaging the berm.

Swales can also overspill into ponds, which enable you to retain rainwater for future use. When you're building swales, look out for natural depressions, which might be good places to use as retention ponds and put spillways in the right place to fill them. You might also use a dam to create a small pond—though if you do, make sure your dam is strong enough and has an overspill mechanism just like the swale.

WATER HARVESTING

Harvesting rainwater from the roof of your house is an easy way to improve your garden's resilience (check out the legal position though, as some states have rules on how much water you can collect and store; others encourage rainwater harvesting).

The size of tank you need depends largely on climate. If rain is frequent throughout the year, get small tanks that can be used and refilled several times a year—maybe three fifty-gallon tanks. On the other hand, if you have a single rainy season followed by a dry season, you may need a much larger tank (e.g., 530 gallons).

So work out your potential catchment before you decide on tanks. Or, if money is short, you can start with one tank and then link more in series, 'daisy-chaining' them together.

Another way of thinking is to ask, how much water does your garden need? One estimate is that a thousand square feet of garden needs a hundred gallons of water a day. If you can collect vastly more water than you need, remember permaculture's value of 'fair shares'—reach out to neighbors who may need more water than their catchment area allows them to harvest.

It's important to keep your gutters clean and cover the tank inlet with a screen to filter out leaves, grit and wind-blown dirt, as well as keeping mosquitoes out. A first flush diverter can also help—a slow drip pipe that will take the first few pints and divert it; this will inevitably be the water with most of the dirt in.

Make sure your tank is opaque or else covered by a tarp. Light getting in can cause algae to multiply. It's helpful to elevate the tank on a platform, so you can take advantage of head pressure rather than using a pump. And if you have a freeze in winter, empty your tanks and pipework, so they don't burst.

Calculate how much your roof will deliver:

If it's 50 by 20 feet, convert that to inches (600 x 240)

Multiply that by one inch of rain to get cubic inches (144,000 cubic inches)

Then divide by 231 to get the number of gallons (144,000 / 231 = 623 gallons)

If you just have a single fifty-gallon tank, you've lost most of it!

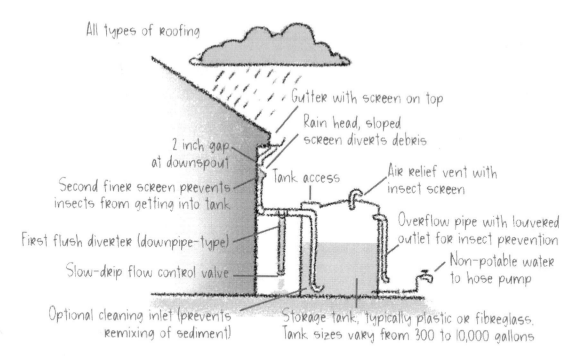

AN EXAMPLE OF CATCHING RAINWATER
TAKEN FROM WWW.BCTRIBUNE.COM

GREYWATER HARVESTING

Greywater harvesting involves reusing water from your washing machine, bathroom sink, bath, shower and kitchen (the toilet is *never* included). Conventionally, this water is seen as a waste product, but it's actually a recyclable, useful resource. It's not just water; it's also nutrients: all the bits of food, skin (yes, I know that's gross), soap and dirt are broken down and made into fertilizer.

But because of the organic content, greywater needs to be treated unless you're only using it to irrigate fruit trees or bushes. It also needs to be used fast; the longer it's left, the worse it smells. Storing it lets all kinds of bacteria multiply and that can also be a health hazard and it should never be used after two days of storage.

The Indian Institute of Technology, Kanpur, reckons that 'fertigation' (fertilizing irrigation) of eucalyptus trees with greywater created biomass of thirty-four kilos per tree, compared to twenty-six kilos with regular irrigation.

The easiest way to use greywater is the way my granddad did; just take the basin of shaving water out and throw it on the garden! (Mind you, I made a point of never eating his lettuce, though his tomatoes were great). It's also easy to use your washing machine's internal pump to pump the water right out to your garden. But remember, this should only go to your trees and bushes, not on the vegetable bed.

Slightly more plumbing is involved in fitting a diverter valve. If you're running greywater through an irrigation system, use 1 inch tubing for the main run with half-inch branches—greywater can be lumpy, so the tubes need to be big. The water shouldn't run at more than a two percent slope—a quarter-inch drop for every foot traveled.

You can take the end of the pipe to a mulched basin, that is, an excavated pit full of mulch, where the water will be incorporated into the soil and mulch and help to irrigate large trees. Because you have to dig pits, this is a bit more time-consuming, but again, cheap in dollar terms. This is best suited to trees and bushes, not smaller plants. Fruit trees don't like salt—a problem if you use a dishwasher, which uses salt as a scourer. But otherwise, they love greywater and thrive on it.

To make this work, use environmentally friendly cleaning products. Above all, don't put chlorine, bleach, or detergents into the system, as they can be toxic to plants.

If you want to clean the water up more, you can construct a wetland, passing the water through a bark mulch filter, then into a reed-bed. A plastic-lined trench filled with pea gravel, with aquatic plants like phragmites reeds, juncus effusus, yellow flag iris and bulrushes (*Typha latifolia*), can do the job. Make sure the water doesn't come above the gravel, or mosquitoes can breed in it.

If you have a freeze or so much rain you don't need the greywater, use your diverter valve to switch back to the mains water system.

You can integrate this system into other systems, for instance, feeding a frog pond which then overflows into a series of swales, with a dam at the bottom to create another pond.

PONDS

Ponds are more natural and cuter looking than tanks. For really large quantities of water, they're also cheaper. Storage tanks at the top of your plot and swales on a slope can feed water through your landscape down to ponds at the bottom, which gather the water before it flows away.

They're great for wildlife. Remember to make the sides shallow so animals can climb out easily. If the sides do seem a little steep, some people construct a simple ramp out of wood or a piece of slate for the same purpose.

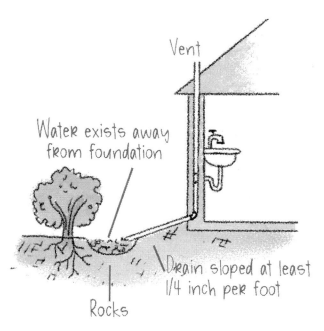

GREYWATER SYSTEM
TAKEN FROM WWW.WALDENEFFECT.ORG

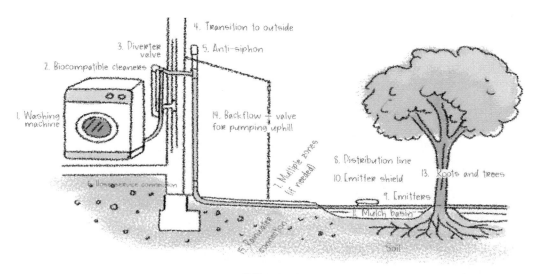

GREYWATER SYSTEM
TAKEN FROM WWW.OASISDESIGN.NET

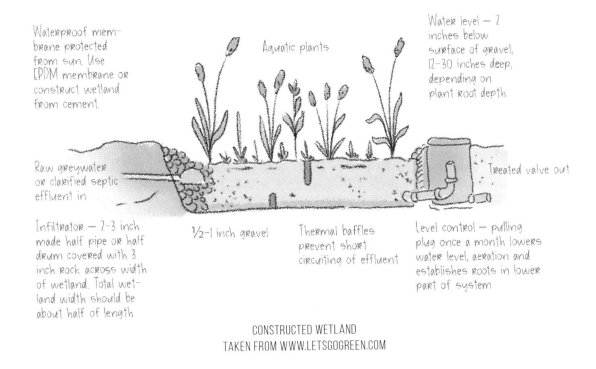

CONSTRUCTED WETLAND
TAKEN FROM WWW.LETSGOGREEN.COM

Tips for using greywater

Become aware of the codes in your area before installing

Use it fast! You can store rainwater, but not greywater

Good for fruit trees and bushes and non-food plants

Be careful not to use water with detergents, chlorine, bleach

If the ground freezes in winter, a greywater system may need to be diverted to the sewer

Water the roots or well-mulched soil, not the leaves or plant directly

Ensure you have a diverter valve so you can switch it back to the regular drains

Use mulch basins in an irrigation system to water trees

Use large diameter pipes (at least 1 inch)

Use a reed-bed system if you want to clean the water up

You can create a pond either by digging a hollow, or by using a dam to hold the water back, or by doing a bit of both. Damming is less time-consuming.

A *gully or key-point pond* is easy to make. Find the key point of the slope, that is, where it changes from a concave to a convex profile. The contour line for your pond's water level runs through that key point. That means your pond utilizes the concave slope of the valley to create a 'bowl' shape, which is the most efficient way of storing the maximum amount of water.

Usually, these ponds store water for irrigation of land below, which is released by a large pipe going through the dam's wall.

A *saddle pond* is formed by a dip in a ridge. You could create one at the top of your garden; it will probably be quite small, but your irrigation will work entirely on gravity.

Contour ponds can be built along a hill where the terrain flattens out to capture runoff coming down the hill. However, they can be expensive to build.

All ponds need spillways. It's smart to join the entire system up, so each pond runs into the next.

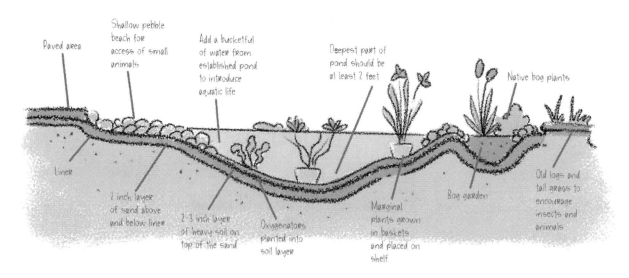

POND DESIGN TO CAPTURE WATER

EXERCISES

Looking and walking

Where is the water on your patch? Where is squishy ground, where is hard? Are there places which are seeping with water? Do you have any springs?

Where are the contours? Walk along them, zig-zag down slopes, try to get a feel for how steep your slopes are. Spend a little time really getting to know your plot.

When you've done this, do a little desk exercise; find out about local wells. In some places, there are still some old holy wells only a little bit below the normal ground level—that tells me the water is close to the surface. In other areas, you might find that farmers have to use fifty-foot-deep boreholes to reach the water.

How many ways?

How many ways can you think of to use and retain water in your plot? You may have a flat area that can't use swales in quite a dry area, but what about trees in pits, stone mulches, using your greywater, creating shade to stop evaporation, sheet mulching, rainwater harvesting? Draw some of your ideas on the map. How do they fit together?

Why water is important

I want you to experience the scarcity and importance of water. If it's hot today, go out for a long run or walk. If not, switch the heating full on and do some exercise in your room till you feel really thirsty.

Go and get a glass. Slowly, pour water into it.

Don't drink it.

Put it in front of you and think about it. What thoughts are coming into your head? Are you really feeling thirsty? Think about how many seeds you could germinate with that glass of water. Think how many birds it would take to drink it, drop by drop. Think how easily it would be absorbed into the soil if you dropped it…

…are you really thirsty yet?

Look at the water. Experience it; clear, cool. Maybe there are little drops of condensation on the glass.

How thirsty are you?

Play it out as long as you can and then, very slowly, take a first sip of water. Savor it.

It may sound like a silly game, but it's a way to appreciate just how rare and wonderful the element of water really is. You won't take it for granted ever again.

5

THE WONDER OF SOIL

Conventional gardening is all about plants. You decide on the plants first, then you fix the soil up if you need to. But permaculture is about getting the best out of what nature can provide. And a huge part of the natural ecosystem is the soil and all the life that it contains.

So let's take a good, deep look at soil. Because next to water, it's the most important thing in your permaculture garden.

WHAT IS SOIL?

Simply put, soil forms the upper layer of the earth's crust. Suppose you were to dig a deep hole; typically, you would see different horizontal layers (horizons) with different textures and colors. There would be a lot of organic matter at the top, then topsoil, then subsoil, and eventually, you'd get to rock.

Another way of looking at soil is that it's a mixture of organic materials, minerals, water, and gas. So you could look at what trace chemicals and nutrients are in it—fertility stored in the soil, as well as carbon and water.

Or you could look at the structure of the soil, the arrangement of the solids and spaces. Well-structured soil is like puffy foam, with plenty of space between the solids to let air move and water drain. Poorly structured soil is compressed, like rolled-out pastry.

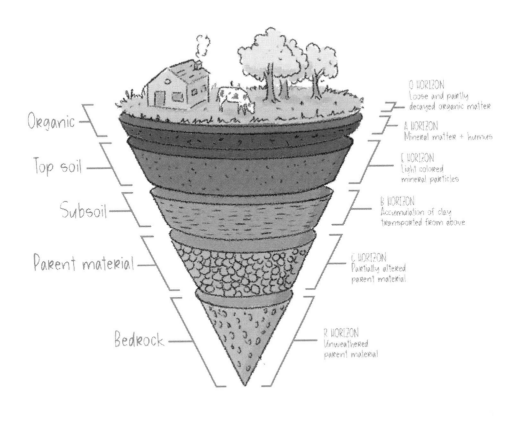

LAYERS OF SOIL

Yet another way of looking at the soil is to look at what lives in it. It's stuffed full of microorganisms and other life, from tiny microbes to fungi—whose roots (mycelium) form a network within the soil—all the way up to earthworms. There are more living organisms in one teaspoon than there are people on the planet.

Most importantly, from a permaculture point of view, soil is active and dynamic. It absorbs water, it transforms organic materials, it stores carbon.

Lynn Kime of the University of Pennsylvania writes: "The soil is alive. In just one teaspoon of agricultural soil, there can be one hundred million to one billion bacteria, six to nine feet of fungal strands put end to end, several thousand flagellates and amoeba, one to several hundred ciliates, hundreds of nematodes, up to one hundred tiny soil insects, and five or more earthworms."

It's a miniature world, and without it, we could not exist.

So first of all, you need to find out what kind of soil you have and what state it's in. There are three main kinds of soil: clay, sand, and silt. Most soils have these ingredients in varying combinations. Anyone who works in traditional earth building knows the 'hand test'—grab a fistful of soil and rub it. If it's gritty, it's sand; if it's floury, it's silt; if it's greasy and slippery, it's clay. In addition to these three, peaty and chalky soils are sometimes considered distinct classifications, but these types are rare.

Press the soil into a ball. Clay will compress and hang together. Sand completely falls apart. And silt will hang together, but not as well as clay. If you roll a clay ball between your hands, you can make a sausage; if you do that to silt, it will crack up.

Loam, which is the best soil for most plants, is a mix of sand, silt, and clay. If you can feel the grit, yet it's still sticky and a bit moldable, that's what you've got.

But that's still just bits of rock ground up very small. For soil to live, it needs to incorporate organic matter. That will make it more granular and crumb-like so that it can absorb water and air can pass through it, and so that plant roots and earthworms can make their way through it easily. A good structure has a balance that is approximately fifty percent solids, twenty-five percent air, and twenty-five percent water—in other words, half 'stuff' and half 'space.'

Soil aggregation comes about by bacteria and roots, together, producing sticky substances that bind the soil particles together. Fungi and tiny root hairs wrap the particles up into balls, or 'aggregates.' That's where the nice, crumbly, soft texture of good soil comes from. The organic percentage of soil might only amount to ten percent of the total weight, but it's absolutely vital and is composed of animals, microbes, worms, fallen leaves, dying roots, and decomposed residue that has become humus.

Organic materials produce energy when they decay—yet another instance of how soil is organic. That allows the soil to hold more water; it also recycles carbon. And better soil supports biodiversity—both the life *in* the soil and the life that grows *out* of the soil.

You'll know if your soil is good: you'll see plants growing well. You can probably tell if it's depleted—if it has no nutrients

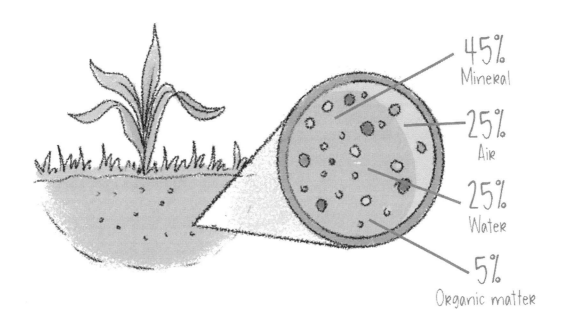

COMPOSITION OF SOIL

left—because whatever's growing isn't doing very well at all.

But you might also have problems with compaction. If you squash soil up too much, it won't drain. For instance, water stays longer in tractor tracks than in the furrows of a field, because the heavy tractor compacted the earth. If you have places that water tends to lie on the surface, that may be the reason.

So how can we change the soil?
Well, you can add a bit of what's missing, like a bit of sand in silty/clay soil. But there are more efficient ways:

Decomposition is a process in which complex biological molecules are broken down into their elements. Earthworms, slugs, and snails break a leaf down into smaller pieces. They shred and chew, pull and tear. That increases surface area. Then the bacteria and fungi get to work, breaking it down into its simple compounds. They secrete enzymes that digest the material. Fungi send out *hyphae* (a kind of root) that break down the material—whitish growth that you'll see if you turn back a layer of leaves.

In the right conditions, decomposition will be fast. It needs to be warm, moist and aerated. If the soil is waterlogged, there's no air in it, and the microbes can't do their work—one reason that 'bog bodies' have been found in peat bogs in Northern Europe. One of the most famous bog bodies is Tollund Man, found in a bog in Denmark in 1950. He was so well preserved that you can still see his wrinkles and slight stubble on his chin, and in 1976 Danish police successfully took his fingerprints.

Organic materials break down into humus (organic matter that can't decay any further). The decay process also releases minerals. Trees are very effective at extracting minerals from the ground through their extensive root systems. They suck up calcium, magnesium, phosphorus, potassium, and nitrogen. They are transported up the trunk, up the branches, to the leaves; when the leaves fall, they're full of those minerals, which are released as the leaves decompose.

Humus holds nutrients. They will leach out much less than in depleted soils. Heavy rainfall can leach nutrients and minerals down to depths where plants can't reach (if you've ever sucked all the orange out of a Popsicle, leaving just white ice, that's what's happening here). But if you've got plenty of humus, the sticky stuff that humus uses to aggregate particles of soil also attaches nutrients to the humus through a process called 'chelation.' Once the humus has locked on to the nutrient, that nutrient can't be leached away.

Humus holds water like a sponge (it can hold ninety percent of its own mass in water). It can also hold oxygen, which is needed for root development, and it feeds microbes in the soil. It also helps insulate soil and protect it from extremes of temperature, and it can even correct soil that's too acidic or too alkaline. Clay will dry up to become quite solid. Add enough humus, and it will stay moldable. Sand on its own can't hold water at all or nutrients. But it can if you add enough humus.

Humus gives soil the right structure, making it not so dense that roots can't penetrate it and not so unstable that plants can't stay upright. Humus also renders soil less vulnerable to erosion. Humus, basically, is what makes soil *work*.

THE BALANCE OF NUTRIENTS

The balance of nutrients in the soil is important for your plants' health. Just like you, they need a balanced diet.

For instance, if plants don't get enough nitrogen, they won't grow properly; their leaves may go an unhealthy yellowish color, and they may have tiny fruits. On the other

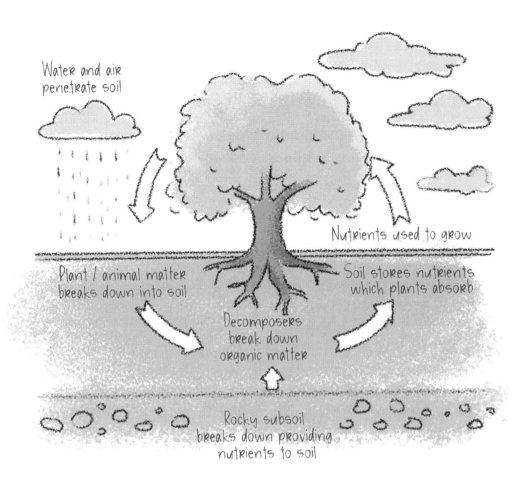

NUTRIENT CYCLE

hand, if they get too much nitrogen, they will produce plenty of biomass but not enough root structure (too much nitrogen content can also make them unhealthy eating for you or for animals, though only poisonous in very extreme cases).

Many farmers add nitrogen fertilizer to the soil, but water run-off can take it into streams and rivers where it harms wildlife, or it can leach into aquifers. Excess nitrogen in water can cause eutrophication, excessive algal or phytoplankton growth. Eventually, as the algae or phytoplankton start dying and rotting, their decomposition takes oxygen out of the water, leaving not enough oxygen for fish and other water life, thus creating 'dead zones.'

Some of the natural soil supplements you might use, like straw or sawdust, are low

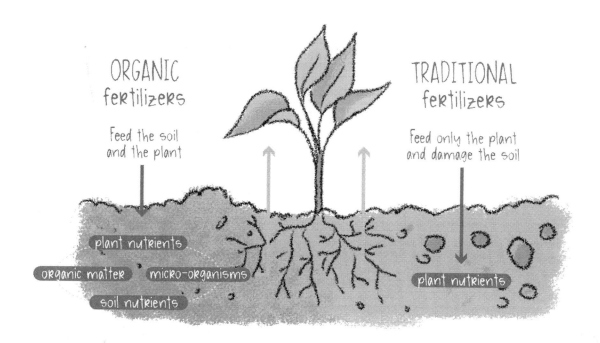

ORGANIC VS. TRADITIONAL FERTILIZER

in nitrogen. So you need to add extra supplements to them that give your plants the nitrogen they need; blood-meal, for instance, is a good addition.

As well as nitrogen, carbon is another element that's important for soil health. The soil stores carbon (as soil organic carbon—SOC), which mainly enters the soil through the decomposition of organic remains and through root exudates and microbes at work. Organic matter in the soil is about fifty-eight percent carbon. Soil organic carbon is a key function of soil fertility; it releases nutrients for plant growth.

(We'll talk about carbon and nitrogen again when it comes to composting, by the way).

Carbon sequestration—storing more carbon in the soil and releasing less to the atmosphere—is a key element in the fight against climate change. Adding organic supplements to the soil, such as green manures and crop residuals ('waste' if you're not a permie), can help increase the amount of carbon held there. Additionally, controlling erosion by using cover crops and reducing tillage or plowing helps to reduce the loss of SOC. So all these permaculture methods have the benefit of reducing greenhouse gases as well as making your garden more fertile—a double whammy of benefits!

Other nutrients the soil needs include:

Phosphorus, which is a major structural component of DNA. Some plants take it up through their roots by a symbiotic relationship with mycorrhizal fungi that help them process it. Deficiency can cause very intense green or reddening of leaves.

Potassium, which helps to form carbohydrates and proteins, is important for fruit formation. Potash—a mined or manufactured salt that contains water-soluble potassium—is the main fertilizer for improving potassium, but there are also very small amounts in manure. You can see potassium deficiency through yellowing leaves, die-back, or stunted growth.

Sulfur, *calcium*, and *magnesium* are also vital because, without them, plants can't perform photosynthesis.

Soil can lose these minerals through leaching, often through excessive irrigation or through soil erosion, when topsoil is stripped by wind or water. Burning vegetation also destroys organic matter in the soil, for instance, in a forest fire.

Big agriculture has to take a lot of blame. Mono-cropping and continuous cropping (of annual plants) both reduce nutrients. Mono-cropping uses only some of the nutrients, but it will use those up completely, leaving the soil out of balance. Continuous cropping eventually uses up the nutrients unless crops are rotated, or the land is left fallow from time to time.

PUTTING THE LIFE BACK INTO THE SOIL

"Thou shalt not dig" is one of the first commandments of permaculture. Most people think that it's a farmer's job to plow and a gardener's job to dig. Huge amounts have been written about the practice of double-digging; millions of rotavators are sold every year. Permaculture says: "You guys are crazy! Why are you doing something that makes the soil worse?"

The idea of plowing or digging is to change the structure of the soil and to aerate, loosen and mix in organic matter. But it doesn't really work. It can compact the earth, and by leaving the soil bare, it makes it vulnerable to erosion by wind or water. When organic matter is exposed to the air, it oxidizes, destroying the nutrients. And tilling uses huge amounts of energy—yours, with a spade, or from fossil fuels with a tractor.

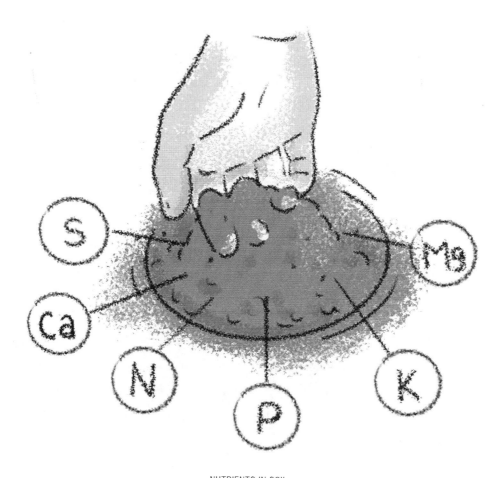

NUTRIENTS IN SOIL

Moving to no-till for commercial farmers can take three or four years of lower yields before things stabilize. However, life is easier for a gardener. What you want is *tilth*—a lovely word, which means the physical condition of good soil that's well-aerated, well-drained, crumbly, and that can retain moisture. That's what we want our soil to be like all the time.

And you can get tilth without tilling! Where you want to open the soil up, use a garden fork and just wiggle it back and forth gently to make holes that will aerate the soil and allow water (and roots) to penetrate.

To improve the soil dramatically in just a year, try sheet mulching or 'lasagna mulch.' If you've got weeds, just chop them down and leave them where they are. Open the soil a little with a fork. And if you have any big plants—trees or big shrubs—to plant, do it first; it's not so easy afterward.

Regular mulching

You need to add organic matter every year because every year, your plants use up some of the nutrients in the soil. Typically, this is a job for the fall. You simply lay down a layer of organic material on top of the ground. Just like the sheet mulch, it encourages good tilth by retaining moisture and nutrients and helps create a barrier to weeds.

Some mulches are attractive; pine needles and bark chips are elegant mulches, as is leaf mold. You can also use a green mulch or chop'n'drop, taking advantage of deep-taproot plants like dandelion, rhubarb, or comfrey that pull up nutrients from deep in the earth. Cut the plants at the base and leave the roots intact. Or you can use straw.

You can also use a 'living mulch'—plants planted underneath your primary crop, particularly if they're companion plants or you might want to pick them together, like basil and tomatoes. Clover and sweet alyssum are particularly good, but you can also use calendula, borage, thyme, or nasturtiums.

There's one thing you must never use, though—diseased plant material. If your tomatoes get blight, rip the vines out and burn them or put them in the trash. You don't want to pass the disease on (some people use this material in a very hot compost pile, since heat should destroy the disease. I don't think it's worth taking the risk).

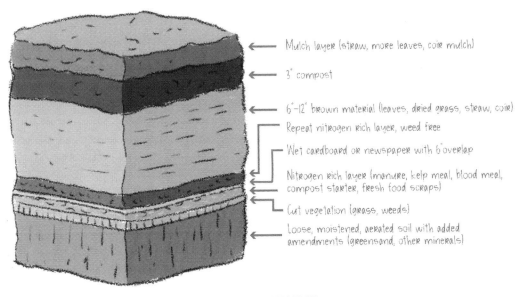

LAYERS OF SHEET MULCH
TAKEN FROM WWW.GARDENERS.COM

How to make sheet mulch

The day before you lay the mulch, give the whole area a real soak with water. The next day, cover the entire ground with a layer of cardboard, or you could use old newspapers. Make sure you overlap the edges, then soak the cardboard.

On top of that, put organic matter such as food scraps, grass clippings, manure, straw, or shredded leaves. You want this to provide nitrogen, so remember this needs to be a green layer. You could even use chopped comfrey or the green parts of prunings from the garden if you've been chopping a lot of stuff down. Add any soil amendments you want, like blood-meal, kelp, or lime.

Then cover that with a second layer of 'brown.' This could be straw, dead leaves, bark chippings, or wood shavings, or more cardboard. Again, give it a good soak; water it until it's really saturated and heavy.

On top of this, scatter a nice layer of a couple of inches of compost. This is what you'll plant seeds into directly, so the quality is important. Use good compost; it's tempting to use topsoil from elsewhere on your plot, but you run the risk of introducing weeds.

Cover over the compost with another carbon layer such as bark chips; this helps lock in all that moisture and it also stops weeds getting started in the compost. Another soaking will help soften the cardboard, and it will also attract earthworms. If you want to, you could even now sow green manure like clover, which will help break down the materials in the bed and will fix nitrogen. Note that not all green manures will fix nitrogen, only leguminous ones, so clover should preferably be chosen.

Sheet-mulching properly needs a lot of organic matter. A fifty-square-foot patch will need an entire pickup truck full of straw (or other mulch). You might want to start small, either centering your mulch on a fruit tree or planting really close to the house to make your first vegetable bed.

There are two crucial parts to success. First, make the layers thick. Second, get them really, really soaked with water. Other than that, sheet mulching is quite forgiving. If you don't have exactly the right ingredients for each layer, improvise.

If you can leave it for three months, you'll have marvelous soil, so it's a great thing to do in the autumn and leave it with an over-wintering crop of green manure. Then it will be ready for spring planting. Or you can plant seedlings by just digging a hole through the cardboard.

How does it work? The cardboard does two things. First of all, it deprives the weed seeds underneath of light, so if perennial weeds try to grow, they won't be able to force their way through the cardboard. Secondly, as it rots down, it will attract earthworms and other soil organisms to break down the organic materials in the mulch.

By the way, if you're doing permaculture on a shoestring, look for free supplies; try looking for supermarkets or other outlets that have cardboard boxes going begging or a farm that has a waste pile of used straw.

Non-living mulch	Benefits	Notes
Grass clippings	Provides nitrogen	Decomposes quickly. Layer thinly to prevent matting and odors. Reapply regularly. Check no fertilizer has been used
Leaves	Weed suppression Water retention Keeps soil cool Provides trace nutrients	Ideally shred the leaves. Apply 2 inches deep to avoid them becoming matted. Ideally use leaves a few years old or leaves that have been piled up in a bin for a year. Don't use walnut leaves. Mix oak with other leaves. Can be composted for 2-3 years
Pine needles	Provides nitrogen Weed suppression	Hardy mulch that only needs replacing once a year. Soil remains well aerated. For acid-loving plants
Fallen branches	Weed suppression	Break into smaller pieces
Twigs	Weed suppression	Break into smaller pieces
Bark Wood chips Sawdust	Weed suppression	Low in nitrogen, therefore use for weed suppression. Avoid if wood has been treated. Can often be delivered free from local tree services. Perfect for perennials but should never make contact with the stems/trunks. Good for pathways. Don't mix them in, just lay them on top. Break down in 2-3 years
Straw	Weed suppression Provides nitrogen (if contains manure)	Breaks down quickly, wet it to prevent blowing away. Don't use hay unless you hot compost it first
Newspaper	Weed suppression	Wet before laying and keep damp
Cardboard	Weed suppression	Similar to newspaper but more effective at suppressing weeds. Use corrugated and avoid any coated in plastic. Make sure pieces overlap and cover with compost, straw and/or bark
Grain hulls	Increases water percolation	By-product of buckwheat, cocoa, rice, cottonseed and other grain crops. Slow to decompose so don't add nutrients. Combine with other mulches to prevent it blowing away

Non-living mulch	Benefits	Notes
Mineral mulches	Increases water percolation Weed suppression	Stone mulches are slow to break down so don't provide nutrients
Manure	High in nitrogen Slow release fertilizer	Don't use dog or cat poop. Spread 6 months before planting or in the fall. Needs to be composted first or it can burn plants. Top with leaf mold or wood chips. Try and source from organic farms
Worm compost	Provides nutrients to the soil	Any time
Greensand	Slow release glauconite	High in trace minerals and potassium
Compost	Slow release fertilizer Good soil conditioner	Make sure it is completely decomposed. Improves drainage, moisture retention Add 3-4 inches before planting. Good around trees (don't let it touch the trunk). Scatter and crumble and reapply regularly

Living mulch	Benefits	Notes
Annuals: Nasturtium Sweet alyssum Calendula Borage	Weed suppression Retain moisture Reduce soil erosion Habitat for beneficial insects	Use in the vegetable garden
Perennials: Comfrey Rhubarb Thyme Oregano White clover	Weed suppression Retain moisture Reduce soil erosion Habitat for beneficial insects Green manure Chop 'n' drop	Use under perennial crops such as fruit trees, comfrey leaves are steeped in liquid; stinky but full of nutrients

Cover crops

Remember, 'one function, many elements'? *Cover crops* act as a mulch, retain moisture, suppress weeds, can provide nutrients and organic matter, improve drainage, aerate the soil, and can give you food and flowers too.

In a conventional garden, when you've got nothing productive to grow in the vegetable

Green mulch	Benefits	Notes
Rhubarb leaves Chickweed Parsley Chives Purslane Comfrey Jerusalem artichoke Yarrow Dandelion	Weed suppression Retain moisture Reduce soil erosion Habitat for beneficial insects	Leave roots of weeds intact to feed the soil and cut at the base. Chop into smaller pieces and lay on top of the soil beneath but not touching the crops. Can add a layer of leaf/wood mulch on top for aesthetics

bed, you leave it uncultivated. In the permaculture garden, you plant a cover crop. Some of the favorites are buckwheat, cowpeas, daikon radish, mustard, and sunflowers—how can you resist a bit of Van Gogh in your vegetable garden? When spring comes, and you want to use the space for growing food, just cut your cover crop down and use it as green manure.

There's another small benefit to cover crops, which is that if you are smart, you can use them as part of crop rotation. So, for instance, if you've been growing tomatoes in part of the bed, put in a nitrogen fixer like alfalfa or clover. Changing between plant families (brassicas, legumes, potato/tomato and root crops) helps to break pest and disease cycles.

Comfrey is permaculture's number one hardworking plant. It's a perennial, grows from zones 3-9, is very easy to grow, has deep roots making it a dynamic accumulator, makes good mulch, and can provide fertilizing 'tea' and forage for animals or chickens. There are two types of comfrey, the self-seeding true or common comfrey and the infertile Bocking 14 or Russian comfrey. The latter is infertile, so it won't self-seed—if you have a smaller plot, this can prevent the potential issue of the comfrey becoming invasive.

Kale will grow over winter in zones 6-9 and is edible (raw or cooked), a 'cut and come again' crop, with lots of vitamin C and antioxidants. Cut it down at the end of the winter and the leaves will return nitrogen to the soil. Note though, that the woody stem of mature kale needs to be chopped up very well for it to biodegrade within a reasonable timeframe.

Not all cover crops are for winter. Sometimes you just have a patch that you want to cover between scheduled crops. Purslane (summer variety), a creeping salad herb full of omega-3, is a good option; dewberry (growing in zones 3-10), a hardy, self-propagating and low maintenance berry, will grow in part shade, but watch out for its thorns. Nasturtiums (zones 8-11) will flower beautifully, distract aphids away from other crops and produce seeds you can pickle; the flowers and leaves are edible too.

There are also a couple of real food crops that do a great job of food cover; squashes and sweet potatoes. Both will grow in zones 3-10, though squashes may need protection from frost; their leaves sprawl across the ground and cover it very effectively. In addition, squashes' prickly stems and leaves discourage pests, and sweet potatoes' roots break up compacted soil.

Did I mention that they're both delicious?

Plenty of herbs will also make good cover—oregano, thyme, marjoram and particularly creeping varieties. They'll help deter pests. But they may need protective mulching over the winter in colder zones.

Annual cool weather cover crops						
Crop	N fixer	Soil preference	Tolerates poor soil	Height	Insectary	Comments
Austrian winter pea	•	Heavy		2ft	•	Hardy to 0°
Barley		Loam		2-4ft		Mild winters only
Bell bean	•	Loam	•	3-6ft	•	Opens heavy soil
Blando brome grass		Many	•	2-4ft		Drought tolerant
Clover, alsike	•	Heavy		2ft	•	Can take acid soils
Clover, berseem	•	Many		2ft	•	Hardy to 18°
Clover, crimson	•	Loam	•	18in	•	Hardy to 10°
Clover, red kenland	•	Loam		2ft	•	Short lived perennial
Clover, sweet white	•	Heavy		3-6ft	•	

Annual cool weather cover crops						
Crop	N fixer	Soil preference	Tolerates poor soil	Height	Insectary	Comments
Clover, sweet yellow	•	Loam		3-6ft	•	Drought tolerant
Clover, nitro Persian	•	Many	•	2ft	•	Hardy to 15°
Daikon radish		Most		1-2ft		Nutrient accumulator
Fava bean	•	Many		4-8ft	•	Hardy to 15°
Fescue, zorro		Many		2ft		Mix with legumes
Fenugreek	•	Many		2ft	•	Opens heavy soil
Garbanzo bean	•	Many		3-5ft	•	Slow in cold soils
Mustard		Heavy	•	2-4ft	•	Opens heavy soil
Oats		Many		2-4ft		Mild winters only
Oil seed radish		Many		2-4ft	•	Hardy to 20°
Phacelia		Many	•	2-3ft	•	Hardy to 20°
Rapseed		Loam	•	2-3ft		Opens heavy soil
Rye		Many		2-4ft		
Ryegrass, annual		Many		2-4ft		Mix with legumes
Vetch, common	•	Many		3-6ft	•	Hardy to 0°
Vetch, hairy		Many	•	3-6ft	•	Hardy to -10°
Vetch, purple	•	Many		3-6ft	•	Hardy to 10°
Information taken from 'Gaia's Garden' by Toby Hemenway						

Annual warm weather cover crops						
Crop	N fixer	Soil preference	Tolerates poor soil	Height	Insectary	Comments
Black eyed peas	•	Many		3-4ft	•	Chokes weeds
Buckwheat		Loam		1-3ft	•	Chokes weeds
Cowpeas, red	•	Loam	•	1-2ft	•	Drought resistant
Lablab	•	Many		5-10ft	•	Drought resistant
Pinto beans	•	Loam		2-4ft	•	Drought resistant
Sesbania	•	Many	•	6-8ft	•	Drought resistant
Soybeans	•	Many		2-4ft	•	Mix with non-legumes
Squash		Loam		1ft		Ground cover
Sudan grass		Many	•	6-8ft		Mix with legumes
Sunn hemp	•	Loam	•	3-6ft	•	Tolerates acid soil
Sunflower		Many	•	Up to 15ft	•	Need sun
Sweet potato		Many	•	1ft		Ground cover, nutrient accumulator

Information taken from 'Gaia's Garden' by Toby Hemenway

Perennial cover crops						
Crop	N fixer	Soil preference	Tolerates poor soil	Height	Insectary	Comments
Alfalfa	•	Loam		2-3ft	•	Well limed soil
Birdsfoot trefoil	•	Many	•	3-5ft	•	Drought resistant

Perennial cover crops						
Crop	N fixer	Soil preference	Tolerates poor soil	Height	Insectary	Comments
Chicory		Heavy	•	2-3ft	•	Opens heavy soil
Clover, strawberry	•	Many		1ft	•	Needs moisture
Clover, white dutch	•	Many		6-10in	•	Needs moisture
Clover, white ladino	•	Many		1ft	•	Needs moisture
Clover, white New Zealand	•	Many		1ft	•	Needs moisture
Fescue, creeping red		Many		2-3ft		
Orchard grass		Many		1-2ft		
Ryegrass, perennial		Heavy		2-3ft		
Timothy grass		Heavy		2-3ft		Needs moisture
Information taken from 'Gaia's Garden' by Toby Hemenway						

Chop-n-drop

Nutrient accumulators reach down to grab nutrients with their roots and bring them up above the surface in their leaves. When you chop them down, the nutrients are recycled into the top layer of soil, where they're accessible for other plants.

Comfrey is probably the most popular; yarrow is another good chop-n-drop crop. In fact, just about any plant you've grown as a cover crop can be chopped. For best results, chop the leaves up fairly small and lay them as a good, thick mulch, a couple of inches deep.

Manure

Manure is a great resource for mulching but buying manure and straw can be risky. Some farmers use aminopyralid herbicides such as Grazon on their pastures and the chemical can stay in the manure and will deform your plants.

Unless you know a local farmer who controls all the materials and can tell you hand on heart that he or she doesn't use this stuff, outsourcing your nutrient needs is always a risk. If, on the other hand, you know an honest farmer, even better an organic one, they deserve your support.

If you have your own chickens, ducks or rabbits, or larger livestock, you're good to mulch. In general, when using manure (other than commercially processed products) in the home garden, it is best to allow it to age first for six months to avoid any potential problems. If you use it fresh, it can burn the plants or lead to salt damage.

Composting

Composting is probably the most familiar element of a permaculture garden. Almost everyone is familiar with the concept of composting—rotting down organic matter to create a nutrient-rich, crumbly, black material. A dose of compost can give a real boost to soil fertility.

Adding some compost at the end of the growing season will put back all the nutrients your plants have used. During the winter, rain and earthworms and similar creatures will push and pull the nutrients from the compost further into the soil, so by spring, you have lovely, fertile soil ready to feed another year's growth.

Remember that composting fits really well into the permaculture philosophy of 'closing the loop'—having a no-waste system where everything is recycled. You grow carrots, you put the carrot peelings and the chopped leaves into the compost, along with other stuff, and you use the compost back on the vegetable bed a while later to put fertility back into the soil. No waste. It's beautifully simple.

But composting isn't as easy as some people think it is; there are a few basic rules.

First, you have to get the right proportion of carbon to nitrogen, or as permies say, "brown to green." While the word 'compost' might make you think 'kitchen waste' ('green'), in fact, a compost pile ought to be twenty-five to thirty parts carbon ('brown') for every one part of nitrogen. Don't think about that number though, that's comparing how much of each chemical element you want. Instead, think about equal parts in terms of weight. In terms of volume, the carbon is pretty light, so you're going to want four times as much volume of brown, carbon-rich material as you do green, nitrogen-rich material—a one-inch layer of grass clippings equals a four-inch layer of carbon, roughly.

If a pile is slimy and smelly, it probably needs more carbon—more brown, or else it may need turning (emptying and restacking, in the same bin or another one) to let more oxygen in. For maximum effect, do both at once. Make sure you break the pieces up before they go in the compost, or they'll be too big to break down quickly. If a pile is just not 'working,' not breaking down, then it needs more nitrogen—more green. Nitrogen is the kick-starter, the ignition; if there's not enough of it, then the composting process won't begin. Also, aim for thin layers of green and brown rather than 'chunks' or thicker layers. Then, the nitrogen and carbon elements share the maximum surface area with one another, which will help the speed of decomposition.

Keep a carbon layer on top and it's going to smell better, and also keep the pile covered. You need to keep the compost moist, but you don't want it soaking when it rains and dry the rest of the time—you want to control the moisture level. If you don't, the carbon parts won't rot properly. You need to keep adding water all the time—the pile should be damp, like a wrung-out sponge or wet towel.

Now, this is all work. Remember that permaculture principle about the most impact for the least effort? So I'm going to suggest something that helps make successful composting less work. Have a compost station that includes a big box of carbon (wood chips, twigs, leaves and cardboard that's already chipped or torn up) next to your composting pile. Then, every time you put some kitchen scraps or grass clippings or manure on the pile, cover it over with a scattering of the carbon. It stops the smell, keeps the proportions right and is very little work. Also, site your composting station close to a tap, hose or water tank. Again, it's very little trouble, if it looks a bit dry, just to water it quickly while you're there.

You need to keep your compost pile aerated. Composting is the work of lots of little bacteria and microbes which need oxygen to breathe. If you press down your compost (or it's pressed down under its own weight) and there's no air left in it, the helpful microbes will die, and you'll be left with a smelly, nasty pile of slime.

One thing that helps is to build a little criss-cross of small-diameter branches at the bottom of the pile. That lets air seep in from the bottom. Another thing you'll probably want to do is 'turn' your heap every so often to aerate it. That means you'll need it to be in a bin or bay where you can open one side to fork it about. Or you could incorporate lots of fluffy straw or other carbon that includes lots of air—small twigs, for instance, laid across each other. That will help it rot fast.

Material	Carbon:Nitrogen ratio
Browns = High carbon	**C:N**
Ashes, wood	25:1
Cardboard (shredded)	350:1
Corn stalks	75:1
Corn stover	60:1
Fruit waste	35:1
Leaves (dry)	60:1
Newspaper	800:1
Newspaper, (shredded)	175:1
Peanut shells	35:1
Pine needles	80:1
Ryegrass (flowering)	37:1
Ryegrass (vegetative)	26:1
Sawdust (hardwood)	325:1
Sawdust (rotted)	200:1
Sawdust (softwood)	600:1
Straw, wheat	75:1
Wood chips	400:1
Greens = High Nitrogen	**C:N**
Alfalfa	12:1
Chicken manure	7:1
Clover (flowering)	23:1
Clover (vegetative)	16:1
Coffee grounds	20:1
Cottonseed meal	5:1
Cow manure	18:1
Finished compost	16:1
Fish scraps	4:1
Food waste	20:1

Material	Carbon:Nitrogen ratio
Garden waste	30:1
Grass clippings (dry)	20:1
Grass clippings (fresh)	15:1
Hay/grass mix	25:1
Hay, mature alfalfa	25:1
Hay, young alfalfa	13:1
Horse manure	25:1
Human manure	8:1
Human urine	8:1
Leaves (fresh)	30:1
Manures	15:1
Rotted manure	20:1
Seaweed	19:1
Vegetable scraps	25:1
Weeds	30:1

Information taken from www.planetnatural.com and 'Designing and maintaining your edible landscape – Naturally' by Robert Kourik

You also need your compost heap to be the right size. Ideally, to kill off any weed seeds in the heap, it needs to heat up to 55-65 degrees centigrade (131-149 Fahrenheit). A pile that is too small won't do this. If you haven't got enough material for a heap (at least a yard on each side), you can still make compost, but you may be spreading weeds… so be careful what you put in it. If your heap is much bigger than three cubic yards, you'll be better off starting a second heap.

When you get a heap started, it can help to add some microorganisms that are already working away. So add a little of your last pile of compost, or grab a handful of good forest floor soil. Or just take a pee on the pile (it's natural, after all). Comfrey, nettle and yarrow plants are also good accelerators, as are lawn clippings.

There is no shame at all in buying store-bought compost to get started. But you may also find your municipality has a composting scheme or that other local permaculturists will let you have some compost to get your sheet mulching started (remember 'fair shares' and invite them to enjoy some of your crops later!)

You can buy a compost bin, or you can easily build one. You could just use a heap, though that's not the most efficient way to compost. Or you could use wire fencing to make a big round bin. When you're finished, just knock it over and the compost is ready to use.

You can also use recycled pallet wood or even complete pallets to create a square bin. Leave the front open, so the compost is easy to get at. Real luxury is having three bins; one that you're adding to, one that you move the compost on to once the fill-up bin is full, and one for finished compost.

IMPROVING CLAY SOIL

If you have clay soil, you'll need to spend some time improving it. If you have a slope, put in some swales; organic matter will build up where the water collects and the rest of the land will drain faster.

Aerate the soil; just use your digging fork and, working backward, use it to wiggle the soil about and open it a little, without turning the soil over. After aeration, you can add soil amendments such as compost, green manure like comfrey or chop-n-drop mustard, leaf mold, or worm castings. Mulching clay soil helps improve its structure and adds organic matter. By adding soil amendments just after aeration, you're ensuring the little holes in the soil can take in some of the goodness next time it rains.

After you've done this, you can call on the plants to help. Daikon radish is a great clay-buster with its big roots, up to two feet long (it's also delicious). Note, however, that these are best kept as an autumn or late-winter crop if you want to have them as a crop, too, as they bolt in warm weather. Mustard has a different style, with a huge, fibrous system of smaller roots that work their way through the clay. It also makes a great bio-mulch but chop it before it produces seed. Sunflowers also help, as well as brightening your garden; chop the tops at the end of the harvest but leave the roots in the soil to rot down. Potatoes and other root vegetables can be used too.

Plant cover crops that will root thickly to reduce erosion and cracking soil. Clover, vetch, grass and legume mixes such as rye with oats and fava beans are all good for the soil, while cowpeas do particularly well on clay.

Finally, remember not to walk on the soil and compact it.

Berkeley hot compost method

A slight variation is the Berkeley hot compost method. It does in eighteen days what cold composting does in six months. Because it's fast and hot, it's good for getting started and for a garden where your composting materials might include a lot of weed seeds. It also produces finer compost. But it requires some hard work.

It needs a twenty-five- or thirty-to-one carbon-to-nitrogen ratio, a heap one-yard square and about five feet high. It will get to a temperature of 55-65C/131-149F.

You can measure the carbon ratio really scientifically (and UC Berkeley did), or you can just do 'one bucket of greens and two buckets of browns' in thin layers till you've filled up the whole bin. Add a compost accelerator like comfrey, nettles, or old compost. Then saturate the heap with water and let it rest for four days.

On day five comes some hard work. You turn it inside out. Basically, you chop down each side of the heap and use the outside to build the middle of the pile next door. Then you take what was the inside of the old heap, and you build the 'walls' around the new pile with that. Make sure the heap is still nice and moist, then leave it for a day.

On day seven, measure the temperature in the middle. It should be on target. This is the point at which the heap will be hottest; after day nine, it will gradually cool down. On day seven, turn the heap again, then leave it a day, then on day nine, turn it again.

You can adjust the carbon/nitrogen ratio as you go. If the heap shrinks too quickly, or it's a bit smelly, it has too much nitrogen, so throw in a bit of sawdust every forkful when you're turning it over. If it's not heating up enough, add a handful of blood and bone fertilizer for every forkful to speed it up.

After this, the good news is that you don't need to take the temperature anymore. The bad news is that you carry on turning the heap every second day, so that's hard work on days eleven, thirteen, fifteen and seventeen. On day eighteen, you can rest. You should now have warm, dark brown, fine, nice-smelling compost.

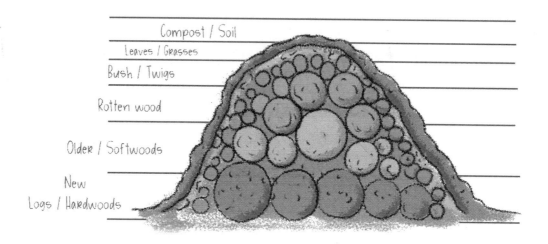

STACKING WITH HUGELKULTUR

Hugelkultur—extreme permaculture

A hugelkultur or 'mound culture' is a way of using big pieces of tree you've chopped down to create a vegetable bed. Dig a big trench and fill it with tree trunk, branches and twigs. You can overfill it a bit (in fact, some people just mound up the wood and don't dig a trench, but if you do this, remember to lay a cardboard sheet mulch to suppress perennial weeds before you add the wood). Then take the turf you cut away and put it root side up over the woody material. Over that, put a layer of leaves and soil, then a layer of half-rotted compost, then a layer of fully rotted compost and then straw. Plant seedlings such as squash and tomatoes through the straw.

As the wood rots, the mound will gradually become lower. The wood retains a huge amount of moisture, which makes hugelkultur a great technique for growing moisture-loving veg in drier climates.

In fact, the hugelkultur mimics the forest floor, where you'll often see vegetation like brambles and fungi growing on rotting, fallen wood. The wood will break down over several years, feeding the nutrients continually into the soil—it's a slow-release method. When wood is kept wet, it becomes soft and breaks down more easily into organic matter. It will also house mycorrhizal fungi which will help stimulate further microbial activity and assist the process of soil aggregation.

One warning: don't ever try to make a hugelkultur next to a swale. The wood is buoyant and that will be a weak point in the berm, which could breach. If you make hugelkulturs on a slope, run them up and down the slope, not across it.

Wonderful worms

Worm castings are a fantastic, rich fertilizer. They have huge amounts of beneficial elements—phosphate, nitrogen and potash; just a few spoonfuls can give a plant a real boost.

So let's get started with worm composting, or as some people like to say (possibly because they think earthworms might scare people), "vermicomposting." You can buy a kit or make your own wormery; keep it somewhere that doesn't freeze, at least over winter.

To make your own wormery, simply get two ten-gallon plastic totes and drill about fifty holes in the bottom of each. You'll want to use a quarter-inch drill bit. Then get a smaller (eighth of an inch) drill bit to make ventilation holes just under the top edge of each tote, as well as about fifty holes in one of the lids (but not the other one).

Put the lid without holes on the floor where you've decided to keep your wormery. Put a brick on each corner and the first tote on top of the bricks. Put shredded paper in the bottom of the tote and moisten it well. Add the worms, not regular earthworms but 'red wrigglers' or Tiger worms/brandling worms, which you can buy over the internet or perhaps grab a few from a friend.

Then add your food scraps. Spread them out evenly. Spread shredded paper out on top and moisten it. Put the ventilation lid (the one with the holes) on top.

All you have to do is keep adding food scraps—only a cupful every week to start with—along with paper, then make sure the bin is moist and wait. If fruit flies or mites appear, you've given the worms too much food to cope with, so slacken off.

What's the second tote for? Once the first bin is full, take the lid off and put the second bin right on top of the compost surface. Then start off the second bin like you did the first, with shredded paper, food scraps and more paper and a good moistening. Over the next couple of months, the worms will migrate through the holes in the bottom of the top bin to where the food is; then, you can then take the bottom bin and use the compost in it.

Worms, by the way, love coffee grounds. They also like most fruit and veg peelings and cores, and even eggshells, but don't give them onions, garlic, or chili peppers. They're not into those flavors.

Trench Composting

If you find the whole idea of starting a compost bin overwhelming, trench composting may be the way to start. You can compost kitchen and garden waste, including weeds with almost no work involved. And it will enrich your soil in as little as one month.

The compost will be entirely invisible, you won't be able to smell it and it can fit almost anywhere in your garden. And the best part? It requires no turning like in a standard compost bin.

How can you get started? It's really easy. Find an empty space in your garden and dig a trench around 12 inches deep, add 4-6 inches of compostable material (this can be kitchen scraps, weeds, leaves etc) and bury them in the soil you dug out of the hole.

There's not much more to it. The compost will enrich the soil and provide your plants with nutrition right where they need it; at the roots. If your local area does not allow traditional composting, this is a great way around it.

FIXES FOR NUTRIENT DEFICIENT SOIL

Fixes for nutrient deficient soil		
Deficiency	Symptoms	Fix
Nitrogen	Yellow or pale leaves Stunted growth	Add coffee grounds Use peas, beans for nitrogen fixing
Phosphorus	Dark edges to leaves Flowers and fruits won't grow or are tiny	Add bone meal
Potassium	Brown/yellowed leaf edges	Bury banana peel an inch deep in the ground
Magnesium	Yellowed leaf edges	Sprinkle Epsom salts
Calcium	Leaves with yellow spots Browned leaf edges	Add crushed eggshell Add calcium carbonate

EXERCISES

Get your hands dirty

Grab a few handfuls of soil in your own plot. If you're visiting other areas, try the soil there too. How is it different?

If you visit someone with a mature permaculture, grab their soil with your hands. They've done this job—they'll understand. How does that good healthy soil feel?

Look for free resources.

First, look around your own plot. Do you have tree prunings, weeds to chop down, drifts of autumn leaves?

Then look around the neighborhood. Abandoned wooden pallets make good bins. Tree surgeons might have chippings and small branches you can take.

What about free labor? If you're making compost or sheet mulching, can you share the labor in your garden with other permaculture practitioners, then go and help them on their plot.

6
MANAGING EXPECTATIONS

How do you feel about the book so far? There's been quite a lot of information to take in very fast, hasn't there? And you may already be feeling a bit overwhelmed. You haven't picked up a spade yet, and it's still been pretty intense.

But remember, in permaculture, we can work in 'chunks.' So you may have a massive, hundred-acre plot, but you can start off with one tree with some redcurrant bushes underneath it and a sheet-mulched area that turns into your first vegetable bed. Put that, together with getting a compost heap going, as your objective for the first year. If you manage to achieve more than that, you'll be happy… as well as tired. In my first year, I dug my top swale and I planted four apple trees, and that was it. I did get some good muscle tone though!

We've already talked about planning what you want to do and how it fits the landscape. Now let's talk about planning your timing and managing your expectations.

There will be hard work ahead. But permaculture is set up to make a lot less work for you in the future. Once you've planted a fruit guild, for instance, you really won't be doing much more in the future than collect the fruit if you've got it right. And 'fair shares' means that if you've got a local permaculture collective or friends, you can get help with some of the heavier work, as long as you share something back.

With most projects, the project manager starts in week one and goes through to, say, week thirty; it doesn't matter whether week one is at Easter, halfway through August, or after the annual budget meeting in January. With a permaculture garden, though, you have to work together with the natural cycle and that means, for a lot of things, with the seasons. For instance, you can't plant a tomato outside before the last frost; you can't plant bare-rooted trees in the middle of the summer; and if you've got clay soil, you can't work it when it's really soggy and wet; you'll have to wait for it to dry out a bit (equally, you can't work it when it's set hard; you have to pick the right time)!

A good way to reconcile your timeline and the seasons and the weather is to make up a long strip of paper and put the seasonal work in at the top—what you *could* do at a particular time. At the bottom, put the structural work in order of priority; if you need to build an access road, that goes in first, with swales next. You may need to slide things along, depending on the weather, but you can see clearly if your plan works, or if you've got one month when everything needs to be done and another when you'll be twiddling your thumbs with nothing constructive that you can do. Some people use whiteboards, or post-it notes, or spreadsheets. Use what works for you.

You might make a multi-year plan. For instance, if you're a busy professional, but you're setting up a permaculture garden in a country home for a decade in the future when you might want to take a break, then you're going to prioritize putting in swales and doing the big stuff—getting things started—while you may decide to leave vegetable gardening till you have the regular time to give it. Or you might say, if you're taking over a garden that's become a jungle, this year we're just going to clear zone one, make keyhole beds and make a compost bin; the rest will have to wait!

On the other hand, if you have a balcony garden, you may only need a one-year plan because most of your plants will be annuals and you'll have everything set up quickly.

Whatever you do, have clear steps laid out so that you know what you're going to be doing and how long it will take.

You may need to manage relationships with neighbors. They probably have expectations too. For instance, if you're in the suburbs, they may expect you to have a front lawn just like everybody else. It's best not to surprise them. Take some time to talk through your plans, explain what you're doing and invite them to have some input. Talk them through the fact that it takes a few years for a permaculture garden to mature.

Some neighbors will just say, "I don't want it to look wild or uncared-for." Others might say, "We have loads of apple trees—why don't you plant some pears and cherries, then we can swap when we have our harvests?" Be particularly sensitive about siting or hiding the stuff that doesn't look too lovely—the compost heap, the water tanks, a big hugelkultur.

Now you have a perfect garden planned on paper, and you have perfect timing planned out for this year and for future years. But you know life isn't perfect. Some things will go wrong. Some things will have to be delayed. Some things may need a bit of rethinking.

It can be easy to feel like giving up. "This stuff is all too difficult," you say to yourself, or "this soil just isn't any good." But you know that isn't true. The great message of permaculture is that given the desire, the will and the knowledge, you can get results out of anywhere. Geoff Lawton has helped reclaim a bit of salty desert in Jordan that is now an oasis; other permaculturists have created gardens in the New Mexico desert.

But be aware things probably won't go smoothly. At some point you'll need to change things, or you'll bust a toe on a rock in the garden, or get fed up tearing yourself on a thorny little bush… so get yourself ready to meet that moment. It'll come—and if you're ready, you'll go and have a little cry, or kick a clod of earth and then you'll go and have a root beer or a cup of tea and think, "Okay, what can I do about that?"

The answers are: change the thing that needs changing, strap your toe up and have a rest—and see a doctor if you think you need to, move that thorny bush to somewhere that it doesn't get in your way, or see if you can find a thornless variety.

And *don't* expect a whole load of fruit and veg the first year. In fact, at the end of the first year, your garden may look like a few twigs and sticks in a sea of mulch. That's okay. Over winter, those plants are going to be delving underground with their root systems, getting some goodness; when spring comes, they'll be ready to grow. Just be patient. Yes, it's hard!

You'll also need to think about some numbers. One big number is how much time you have every week to spend in the garden. You may be able to take a vacation and work together with friends on the first stage, which is fine. After that, though, you're taking time around your day job. The amount of time you can spend might depend on whether it's still light outside when you get home from work, so you'll have a bit more time in summer than in winter.

Remember you need to sleep, eat and relax. You can't budget a hundred percent of your spare time for gardening. You may be ill for a couple of days and not have much energy for a while.

Time is one of the numbers; another is, of course, money. Think about how much you have to spend on the garden. To some extent, time and money are exchangeable. If you're working in a high-paid job, you can hire experts to get your access road and swales made, or you can hire a gardening team. You can buy plants that are more mature and gain a bit of a head start that way. You can buy a load of compost or topsoil to improve your soil in one quick tip of a pickup truck. You can hire a permaculture designer to create your master plan.

On the other hand, if you have a lot of time but not so much money, you'll be doing a lot of work yourself. That might include learning how to propagate different types of plants, so you don't have to buy them from a nursery. It could include sowing tomatoes and squash plants from seed instead of buying the plants.

You could look for free resources like pallet wood; you could use hazel and willow, coppiced or pollarded, to provide you with free sticks for making hurdles, fences and supports, or you could use bamboo the same way (watch out, some bamboos are invasive). Someone with more money would probably buy ready-made fencing. You could plant hedges as fencing, which also have other uses, such as providing habitat for birds and other wildlife. You might decide to daisy-chain together a number of different-sized rainwater tanks you found on Freecycle because buying a new one the size you need would cost too much.

You might also decide to buy standard varieties rather than looking for more expensive, rare plants. But you'll still have to spend something, so budget how much you can afford. Maybe you can only afford to put in two fruit trees this year and you'll put in two next year. Or you could always ask family and friends to buy you trees for your birthday and Christmas presents (one of my friends has an apple tree called 'Tony' after the friend who bought it for her).

Finally, when you're planning, prioritize what's most important for you. We already looked at this; it's usually—though not always—zone one. If your first priority is growing food, start by thinking about what food you eat the most, what food is healthiest or most expensive or difficult to find—that will give you ideas for what you want to plant. Remember that in financial terms, planting things like chili peppers and bell peppers, or squashes, if they're expensive to buy, will save you far more money than planting potatoes and onions. For me, growing rocket and other salad leaves saves me a lot of money at the grocery store. And if you hate rhubarb or Brussels sprouts, don't grow them!

So now you have your design, you have your list, you have your timeline and your costings. You need to make another list, too, of emergent factors (urgent stuff that's got to be done right now) and limiting factors (the factors that stop you from doing some things;

these could be money, winter temperatures, local regulations, or lack of space).

Emergent factors are things like an access that is dangerously slippery when wet, a dam that is giving way, or flooding around the back of the house. These go to the top of the task list straight away.

Limiting factors are dependencies; for instance, they might be tasks that have to be done in a certain season, that have to wait till you have saved up enough money, that have to wait for a professional (some waterworks will require a plumber who can work to code and sign off his work), or that need to wait for another resource to be put in place first. For instance, you can't plant a raised bed till you've made it!

Now, remember that permaculture is about least work for most impact; look for the small things on your list that make the biggest difference. And then think about your work in phases—that's where the middle of that long sheet of paper comes in useful; you can draw a bubble for each phase.

Let's suppose I'm starting a garden in a dry climate. I have mains water but nothing else. What will make the biggest difference right now? Waterworks. So I will put my initial effort into setting up a couple of rainwater tanks, greywater harvesting and maybe putting in the top swale out of four or five that I might want eventually. With the swale, the reason for starting with the top one is that the water from that swale will plume out downwards, so I'll already have recovered that part of the garden before it's time to dig the next one.

Suppose I have a lovely, cleared site in the forest and when I look out the window in the morning, I see deer browsing on the ferns and grasses. What's my first priority? Get a deer-proof fence, plant a row of lavender (they don't like strong smells), or plant thick hedges to keep them out (and remember, your hedge will also provide other uses such as habitat for wildlife). Otherwise, anything I plant is going to be eaten. As hedges can take a while to become established as a barrier, one idea may be to have a temporary fence at first but plant a hedge at the same time and once it has become established, it will provide many ecological benefits as well as keeping the deer out!

Try to do things that have multiple benefits. For instance, if you have a soggy area that floods right outside the back door, then draining that is going to do a lot of things. It's going to make your access to the garden better; it's going to take water away from the house walls where it could lead to rot or subsidence; it's going to improve the soil. It could also flow into a pond lower down.

One city garden I saw in London was transformed by taking down an overgrown sycamore tree. The whole garden was shaded by the tree and got no light at all between the tree, the house and the neighbors' fences. It's not exactly a small effort cutting down a tree, but doing that enabled the whole garden to get sunlight six hours a day, which made the biggest difference to the way the garden could be developed. It also produced loads of chippings and bark for compost, some wood that could be used for making raised beds and old branches that were buried deep in a hugelkultur bed. The impact greatly outweighed the work involved.

DETAILED PLANNING

Make yourself a monthly planner which takes into account how much time you can commit. Plan when to grow seeds indoors and transplant and when to sow or plant outside (remember that your particular climate and microclimates may mean you can plant earlier or need to plant later than it says on the packet).

If you're new to gardening as well as new to permaculture, look for things that are easy to grow and that you like to eat. There is time enough for experimentation once you've gained some expertise. Easy vegetables include salad leaves (which you can pick as soon as they're big enough, just a couple of weeks), radishes (also fast), spring onions, broad beans, bush beans (haricots), kale and chard. Tomatoes, zucchini and squash, are easy if you buy seedlings to plant out. Root vegetables such as sweet potatoes, regular potatoes and Jerusalem artichokes are not hard either.

Don't forget to add your harvest dates into the planner. No-waste gardening means "when it's ready, eat it!" (or at least store or preserve it to eat later, or donate it to a local charity or your friendly neighbor).

You should also think about what to do with a glut, for instance, if you can't eat all the apples you produce. You can share them with friends, neighbors and other permies who've helped you; you can slice the apples and dry the slices, or press them for juice and pasteurize it, or freeze a lot of apple pies… but don't waste them! Prioritize harvesting and processing your harvest over planting new areas of the garden.

Only plant what you can manage. If you are digging a swale and you're doing it on your own, it's better to dig a few meters and plant the berm and the back of the swale before you continue—you don't want it to be taken over by weeds. Leave yourself a little time in hand for rethinks, minor disasters and having a rest some time!

Even if you want a large garden and you've got the land to do it, start small. Get some experience managing a smaller area and expand once you feel confident. This is particularly important if you're completely new to gardening. One way to do this, again, is to share. For instance, several friends in London, UK, got into permaculture together by sharing an allotment garden which they all felt was just too big for any one of them to tackle. Two of them kept that allotment as a share while the third went on to garden a whole allotment herself using what she'd learned, as well as taking on a couple of hives of bees, so her friends got jars of honey for a present every holiday.

While we tend to think of big jobs like digging a swale or planting a whole big vegetable bed, a lot of gardening comes down to small jobs—tucking a couple of orange peels under the mulch on a vegetable bed to add to the mix, for instance, walking and noticing what's growing and what seems to be doing poorly and might need extra nutrients, or picking today's cherry tomatoes and thinning the salad leaves to put on tonight's pizza. A daily ten minutes in the garden can be worth more than a couple of hours once a week.

If you find you don't have enough time, then there are a few shortcuts you can take. For instance, growing from seed takes time. You'll need to have space indoors and you'll need to thin out seedlings, then put them in separate pots ('potting on').

SEEDLINGS

You can cut all this out by the simple (and not terribly expensive) method of buying your tomato, zucchini and squash plants already established. And you can also look for vegetables that you can sow directly in the vegetable bed instead of starting them off indoors. A good little cheat here is to start off your salad leaves in a row with radishes. The radishes grow really quickly and you can pick and eat them quite soon, automatically thinning out the salad leaves in the process. Further thinning occurs quite naturally every time you want to eat salad, so by the time your arugula and lettuce are getting on for normal size, you've done all the thinning and not really done any work! (Did you notice that this method of thinning is another good example of 'one element, many functions')?

Try to grow perennials where you can because then you don't need to sow them again. So with broccoli, you might prefer to grow perennial nine-star broccoli instead of an annual species. Vegetables like asparagus, globe artichoke, chard and kale offer the same kind of eating as spinach or cabbage but can be grown as perennials, as can sorrel. Plenty of herbs are perennials too.

Self-seeding annuals can also save you time. Arugula (rocket), for instance, will sow itself, as will many lettuce varieties; many alliums—onions, chives, garlic—will self-sow (if you're relying on this, remember not to buy an F1 hybrid variety if you want the seed to come true to the same variety as the parent).

Here's a special tip: plant at least two of everything. If you have just one tomato plant and, for whatever reason, it dies, it's a tragedy. If you have two tomato plants and one dies, you still have the other one. It just makes gardening so much less anxiety-inducing. You may also want to consider seed swapping using online forums or local networks that provide you with useful access to other gardeners with surplus plants.

Remember, making mistakes is not a disaster. It doesn't mean you've failed and it doesn't make you, personally, a failure. Making mistakes is the way we learn and grow. Keep a diary of what you're doing in the garden and note what went wrong and what you can learn from it; my entries include:

- I should be more suspicious when the sun comes out too early in the year. Guess what, I planted all my tomatoes the night before an early frost and it killed the lot. And next time, I'll only plant half my seedlings and keep the rest indoors; that would be more sensible.
- Uh-oh, all the tadpoles are gone. *Now* I find out that koi carp will eat tadpoles. I'd better have a separate tadpole pond!
- My compost's gone all slimy. Okay, I need more sawdust and more twigs and brown stuff.

None of these mistakes stopped me from becoming a good gardener. In fact, when you've learned a lesson the hard way, it usually sticks in your mind!

THE WAY OF THE BUSY GARDENER

You may have a busy life; you could be a start-up CEO and father of two and *still* want to get into permaculture and make a small difference to our planet. Is that realistic?

Yes, it is. But you need some help. And because permaculture is about 'fair shares' and 'caring for people' as well as the Earth, you are very probably going to find it.

One option is joining a CSA—Community Supported Agriculture. You get your food from a local farm and you commit to spending some time working on that farm in return. You'll meet like-minded individuals and you'll learn about gardening within a supportive environment. Perhaps best of all for busy people, the hours you work are defined, unlike a garden at home, which can suddenly swallow up all your time.

You may find a community garden you can join, particularly in cities and small towns. A community garden might just give you a single raised bed to look after, which is a very manageable unit in terms of space and time. And as with CSA, you'll meet other people who're going on their own permaculture journeys while learning about gardening in a low-stress, collaborative way.

Or you may find that local gardeners are happy for you to help out for a share of the fruit. One bunch of busy local professionals decided to trade—one ran a vegetable bed, one ran a permaculture orchard, others added different specialties. They all helped each other and at harvest time, all of them shared the produce. Their kids loved working with the other families and getting the occasional sleep-over, so it was ideal for their families, gardens and health.

If, on the other hand, if you've got time but not a garden, you might consider WWOOF—World Wide Opportunities on Organic Farms. You could combine this with international travel if you're so minded as there are WWOOF farms all over the world. You pay a small fee to register and farms will give you bed and board in return for your work (you won't get wages). While young people often use it as a way to get started, there's no upper age limit, so if you're retiring early to set up a homestead, why not WWOOF first? It is also worth mentioning that a lot of the farms are organic gardens, large or small; they're not all large commercial farms. These farms can offer you a more personal experience and may be a good place to start for someone new to permaculture and gardening and wants a friendly introduction.

IF YOU LIVE IN A HOT CLIMATE

Permaculture works anywhere, in almost any climate (perhaps Antarctica isn't going to get a permaculture garden any time soon, but most other zones can). But obviously, there are some differences if you live in a hot climate, whether it's arid land or a sub-tropical rainforest. Most books are written for those who live in temperate zones, so let's just summarize here a few of the differences you'll find if you're in one of the hotter climates.

The basics of permaculture will still be the same: taking into account the natural form of your land, its climate, soil and sectors (constraints); enhancing biodiversity; closing the cycle so that all the resources that come into your garden are productively used and recycled and you waste nothing and using the relationships between the different elements to create resilience and productivity. But some of the details will, necessarily, be different.

First, you'll be able to grow crops all year round. In the cooler part of the year, you can grow traditional 'cold climate' fruits and vegetables, while in summer, you'll be able to grow tropical foods like cassava, taro, ginger, bananas, guavas and mangoes, as well as avocados. Remember, though, that some vegetables from the temperate zone need a 'cold snap' to tell them to wake up in spring; they won't get that in a hot climate (though sticking such seeds in the fridge will get them germinated).

Second, you may have to cope with extreme rainfall, either in a desert climate where the occasional day of torrential rain occurs or in a monsoon climate where you have a distinct rainy season. This can put a lot of pressure on your water system; swales may overflow, so you'll need secondary relief ditches that can take the overflow to pools or tanks.

And third, you don't get a winter break. If you live in a relatively northerly, temperate climate, you probably get two or three months of the year during which, after a little tidying up, you have very little gardening to do and you can get on with something else. In a subtropical climate, forget it—there's *always* something to do.

You'll need to spend more time pampering the soil. Organic materials break down more quickly in the tropics and nutrients are exhausted fast, so you will spend more time on soil improvement than a temperate zone permie. You do have some interesting materials available though, such as sugar cane straw as a mulch. Remember never to mulch dry ground; if you're going to mulch in the middle of an arid zone, then water the ground really well and let it soak in before mulching.

Weeds grow fast. So don't waste your time on them—either eat them, get chickens or ducks to eat them, or turn them into green manure. But pests grow fast in hot climates too, so it may be worth leaving a fringe of weeds around your veggie beds—they can lure the pests away from the food you want to eat. You can also stew up weeds to make a 'weed tea'—stick them into hot water, let the nasty brew stew for a week and then use it on the garden. Throw what's left of the weeds in the compost—look, no waste!

For really bad insect infestations or really tasty fruit that attracts pests, you'll need a barrier. Use netting to keep pests off your raspberries or guavas. This you probably wouldn't bother with in a temperate climate, but then you don't get swarms of locusts in Washington DC or upstate New York (by the way, locusts take to the trees at night and sleep really well. If you knock them down onto a sheet, then bag them, you'll end up with a lot of fertile insect bodies for the compost heap).

If you're looking for inspiration, as far as we know, there's no really good book for permaculture in the heat, but there are some very good videos on Youtube from other permaculture practitioners.

CAN I TAKE A VACATION?

Yes, of course! And if you're practicing permaculture principles, you should actually be able to do so a bit more easily than the intensive 'modern' gardener who may have plants unsuitable for the location, doesn't have swales or ponds, or guilds and has a whole load of bare earth that needs watering. Once your permaculture is set up, it should look after itself.

If you will be away for a couple of weeks, check your mulching; a little extra mulch could help retain moisture. Give the garden a good watering before you mulch and before you go. And if there is fruit or veg that's nearly ready and needs harvesting, get them in—many berries and vegetables can be frozen (it helps to blanch vegetables first in a little boiling water, just two or three minutes, then rinse through with cold water before you freeze them).

It's still worth having a support network, though, as there may be a few things that need looking after while you're away. You might pay a neighbor or simply promise to supply them with fruit and veg, or if you have chickens, let them take the eggs while you're away. Or you might take turns to support another gardener.

It's best if you write down very clearly what are things that will probably need attention; in fact, it might be best to jot them down on a copy of the map of the garden. And for watering, you can use an 'indicator plant' somewhere close to the door—a plant that starts looking sorry for itself and wilting before the other plants do. Lettuce is one of these plants, so your garden-minder can take that as a signal to look at your vegetable bed and see if the other plants have enough.

You could also install an irrigation system, as high-tech or low-tech as you like. Of course, for a lot of your garden swales or your greywater system, you may already be doing this. But if you're using greywater harvesting, you won't be producing any greywater while you're not at home. So use high tech and run the taps from your smartphone! Or you can just set a soaker hose (a hose with small holes intended to drip water into your garden) on a tap and run it under the mulch. You can set a timer, so it only comes on once in the morning and cuts off for the rest of the day. Some people use solar-powered irrigation systems, which consist of a solar panel, battery and pump, which distributes water from a water butt to your plants through drippers. This is a good way to manage your water supply in an economically friendly way.

If you have a small garden or grow pots in containers and you won't be away too long, you can set up a simple system using old PET soda bottles or even milk jugs or Tetra Paks. Just fill up the bottle with water and stick a few small holes in the bottom; it will gradually leak moisture into the earth. It's also a good idea to sink pots and other containers a few inches into the ground, which keeps the roots cool and allows the earth in the pot to suck up moisture from beneath.

Greenhouses are a bit of a problem. They can get too hot if you're away in summer, so leave the ventilators open. Alternatively, you can make grids to put over the greenhouse doors (wooden frames covered in chicken wire) so that you can leave the door open and not have to worry about anything getting in and stealing food. And give all the plants a good soaking before you go; arrange a neighbor, buddy, or an irrigation system to do the watering if you're away for more than a week.

So far, I've talked about what you should do when you go away. But remember, also, to set aside some time in the diary when you come back for checking on the garden, tidying up and generally making sure everything is on track.

Or you could just decide to take your vacation in winter. Go off while it's cold and gray to somewhere the sun is shining, having already packed your garden up in its winter mulch and straw. You'll have nothing to worry about at all!

FINALLY… SOME HELPFUL TIPS ON GETTING STARTED.

As with all walks of life, you can have done the training course, read all the books, watched the videos… but it's different when you actually have to do it for real. A friend of mine is a doctor and she says after seven years of medical training, the first time she sat on her own, in her own office, waiting for a patient to come in, she was in a complete panic. Her mind just went blank. Fortunately, it didn't stay that way long!

So here are a few helpful tips on how to make sure your permaculture garden is a success.

First of all—we've talked about design and principles and planning and how things work—but the most important thing is just to get started. Jump in with both feet. Decide on one thing to do first and do it, whatever it is. Get it done, get it finished. A hugelkultur, a keyhole vegetable garden, an orchard swale—just focus on it and get it done.

You may be designing big, but start small. Divide your overall plan into parts that you can achieve separately. That way, you get a sense of achievement. You might want a mandala garden made up of a dozen keyhole beds, but get started with just one. Build it up, plant it and mulch it. Now you know the process, you have a definite achievement, you know how long this one took. In future, you're going to be able to do just the same, however many times you need to.

The great thing about doing things this way is that you don't end up having things half done. Conventional gardeners often race ahead to dig garden beds and don't get around to planting them. Then the weeds start growing and the gardeners spend all their time weeding and not planting the next bed or feeding the plants; eventually, the weeds win. In a permaculture garden, you build the bed, feed the soil, plant the plants—and then you're ready to go on to the next thing. There's nothing holding you back.

However, within each element, start with the biggest thing, whether that's a swale in the back yard or a big tree in a guild. Then all the smaller things will fit in around it.

Managing your life and your permaculture

Plan for the long term—a permaculture garden can take years to develop

Plan 'chunks' that you can get done rather than trying to do a bit of everything

Prioritize what's most important for you

Try to get the most impact for the least work

Focus on tasks that will provide the most benefit

No garden? Or too busy? Get into WWOOF, community gardens, or garden sharing

Find a support network to cover your vacations

Ensure you have a good relationship with neighbors

Be prepared for things not going to plan and take it easy on yourself when that happens

Make plans around how much time each day/week you have to spend in your garden and keep a planner each month to help you stay organized

Make plans around how much money you have to spend, and this may mean spreading out the work you need to put into your garden

Look to community/neighbors/local support groups to find out where you may be able to find freebies or people to help

Get the help of experts if you feel too overwhelmed or out of your comfort zone

Get comfortable managing a small area of your garden before you expand this out into the whole garden

If you don't like it, don't plant it

Second, remember that permaculture means different things to different people. What does it mean for *you*? Hold on to that thought. You may want a full-scale food forest, or you may want to use some basic permaculture principles in what otherwise looks like a very normal suburban garden. Don't get distracted by seeing other stuff that doesn't help you get *your* garden built and operating. Here are just a few ideas people have put into effect in their permaculture gardens:

- Vertical gardening—growing vine tomatoes, grapes, kiwi fruit, zucchini and gourds up the sides of walls and fences, with other vegetables grown in

their shade and using extensive composting to keep the plants well-fed.
- Succession planting or 'stacking in time'—a small vegetable bed can produce more than you'd think if every time one crop is finished, a new one replaces it. At the end of the year, the new crop is a green manure for overwintering and providing food for next year's plants.
- Square foot gardening—this involves taking a square section of the garden and dividing the length and width by plant spacing needs. You then take the seed and plant spacing numbers to divide up the planting sections and know how many seeds to sow.
- Plant stacking in what used to be a flower bed—planting a fruit tree with fruit bushes in its shade, ringed by vegetables such as pumpkins and zucchini and companion planted with marigolds and herbs. It's the only permaculture element in what otherwise looks like a normal town garden.
- A whole hillside with swales dug into it, planted with fruit trees and bushes with sweet potatoes and other vegetables as an underlayer. Yes, this guy got some help from his friends!
- A balcony with pots of perennial herbs, heavily-mulched pots of tomatoes and a dwarf plum tree, with a wormery to provide compost from kitchen scraps.

Third, whatever the books say about kale, or garlic, or rhubarb, being great plants for permaculture gardens, if you don't like them, don't plant them! If you like strawberries, paw-paw, avocados, whatever, plant those (subject to your climate being suitable). Your garden is meant to be fun; healthy eating happens naturally when you're eating fresh fruit and veg that you actually like.

I hate Brussels sprouts. There are none in my garden. If there were, I'd use them for composting. On the other hand, I love asparagus. And asparagus is a perennial, which fits nicely into a particular couple of places in the garden. So every spring, there's that marvelous moment when the first few spears of asparagus get covered with butter and put in the oven to roast slowly. That's my motivation for gardening!

Fourth, attract buy-in from friends, partners, kids. Instagram your garden and put your garden news on Facebook. Find easy and fun activities for your kids to do (if it involves getting really dirty, they usually enjoy it). Have a garden 'work first, Super Bowl afterward' party for friends.

Square foot gardening – popular vegetables			
Vegetable Type	Plant Spacing Per Square	Vegetable Type	Plant Spacing Per Square
Arugula	4	Oregano	1
Asian Greens	4	Parsley	4
Basil	2-4	Parsnips	9
Beans (bush)	5-9	Peanuts	1
Beets	9	Peas	4-9
Bok Choy (baby)	9	Peppers (Bell)	1
Broccoli	1	Peppers (All Others)	1
Brussel Sprout	1	Potatoes	4
Cabbage	1	Pumpkins	1
Cantaloupe	2 squares per plant	Quinoa	4
Carrots	9-16	Radicchio	2
Cauliflower	1	Radishes	16 or more
Celery	4	Rhubarb	1
Celtuce	2	Rosemary	1
Chives	4	Rutabagas	4
Cilantro	1-9	Sage	1
Collards	1	Scallions	36
Corn	4	Shallots	4
Cucumbers	2	Sorrel	2
Eggplant	1	Spinach	9
Endive	4	Squash	1
Fennel	4	Strawberry	1-4
French Sorrel	4-9	Swiss Chard	4
Garlic	9	Tarragon	1
Green Onions	16	Tomatoes	1
Kale	1	Turnips	9
Kohlrabi	4	Thyme	4
Leeks	9	Wasabi	1
Lettuce (leaf)	4	Watercress	1
Lettuce (head)	2	Watermelon	2 squares per plant
Melons	2 squares per plant	Yams	4
Mint	1-4	Yellow Onion (large)	2-4
Onions (bunching)	9	Zucchini	1

Even if your bestie doesn't actually help with the garden, they'll be a shoulder to cry on when things go wrong—and you may occasionally need that—and when things go right, that's a great excuse for cooking a good meal with your best new veg and fruit and opening a nice bottle of wine.

1 Tomato (Per Square)	1 Tomato (Per Square)	2 Cucumbers (Per Square)	4 Chives (Per Square)
1 Tomato (Per Square)	4 Arugula (Per Square)	1 Oregano (Per Square)	4 Parsley (Per Square)
2 Basil (Per Square)	6 Leaf Lettuce (Per Square)	6 Leaf Lettuce (Per Square)	1 Pepper (Per Square)
9 Green Onions (Per Square)	6 Leaf Lettuce (Per Square)	6 Leaf Lettuce (Per Square)	16 Carrots (Per Square)

SQUARE FOOT GARDENING

EXERCISES

Time yourself

It can be difficult when you start out to know how much time it will take for you to do things. To dig a hundred-yard swale? Plant a tree? Mulch the vegetable bed?

So, time yourself. Dig the first yard of the swale. How long did it take? So a hundred yards will take a hundred times that, plus a few rest periods and cups of coffee.

Mulching the raised bed, say, took half an hour. If you start another raised bed, that will be an hour to mulch the two of them.

Jot these times down in your record book and use them when you're planning your work.

Take photos

Identify five or six places in the garden to take photographs from, which will represent all the different parts of the garden. Every week, take the same six photos.

As your garden develops, you'll be able to look back through the photos to see the progress you've made. It's a real morale-booster to look back and see how things have changed!

Plan for the long term (planning horizon)

Get a big sheet of paper, or stick smaller pieces together till they cover your tabletop—or use a big whiteboard if you have one. Divide it up into the next five years and each year into the four seasons. Now look at the features you want and think about where they need to go and when you should implement them.

Remember to think about time dependencies, like when will you be able to get bare-root trees from the nursery you want to use? When will you have vacation time you can use for extensive works?

This is a really helpful document for you. If you're feeling that nothing's working as you look at a muddy, unattractive, cardboard-mulched garden at the end of year one, look at your plan and see that you've done the things that will give you a much more attractive and productive garden next year! And of course, if you want to change the plan, you can. Perhaps a new baby arrives and suddenly, a nice play area gets added to the list of features. There's nothing stopping you.

7

CHOOSING PLANTS FOR THEIR ROLE IN THE GARDEN

Let's consider how a 'normal,' conventional gardener chooses plants. Let's dip into their stream of consciousness and understand their motivation for their choices.

"Oh look, a catalog from Gardens R Us! Great! Even better, a 10% offer up to the end of the month. Okay, let's look. No, I don't like zinnias… or dahlias… oh this is better, look at these tulips. I love double tulips and those variegated ones, I'll get some of those. Oh, hedging. Skip over those pages… hm, a new blackberry variety with no thorns. Might get that. Wow! Look at these fantastic neon colors? What *are* those plants? I have to have one!"

There's not really much of a plan there other than 'looks nice,' apart from the blackberry (by the way, I really appreciate thornless or less spiky versions of plants for the food garden, though the spikier ones are useful for fencing). So let's look at how a permaculture gardener chooses plants.

For a start, a permie is quite likely to look at the catalog as a symptom of consumer culture. It may end up in the compost. But permaculture choices go like this: "I really need more nitrogen fixers. Let's look up the list… have they got any of those? Good. Okay, I've got three spots where those will be useful. Now for salads, I said this year I was going to try some Chinese vegetables, because I enjoy eating them, so mizuna and pak choi? Oh, the pak choi needs to be in the polytunnel. Yes, I've just got enough space. And I need some green manure seeds, so alfalfa and mustard fill the spot there."

In permaculture, because we see the garden as a system, we look for plants based on their functions and their relationships with other plants and with the soil. So here, N-fixers and green manures are picked because they will contribute to maintaining healthy soil, while the other plants are chosen because they are good to eat. There might even be plants that are good to eat, attract pollinating insects, fix nitrogen and are compostable too—the more functions, the better.

Permaculture is based on the *food forest*, a forest just like a natural one, which is made up of associated species, some of which don't thrive outside the forest ecosystem. For instance, there are animals, like the spotted owl, which need old-growth conifer forest to live in. Some kinds of lichens and mosses also need this particular environment.

Other species are more tolerant and can live in different environments. Some prefer to live in coniferous forests, others in deciduous or mixed woodland. White-tailed deer, for instance, can live anywhere they have access to mature forest as shelter and young undergrowth (or stolen backyard foliage!) as fodder.

Mushrooms, bees who swarm into cleft trees, and all types of berry plants do well in the forest (in Kenya, agroforestry farmers tie little pots into their trees to form houses for bees). Then there are the species that like the forest edge with its partial shade.

Let's look at the role a large tree plays. Its canopy doesn't just provide shade to the plants under it; it also traps dust, absorbs pollution, absorbs carbon dioxide, cools the air and disperses rain so it falls less heavily on the ground and erodes less soil. Because the soil is cooler, the moisture in it will evaporate more slowly. The tree may host micro-habitats exploited by bats, birds, lichen, fungi and insects; a mature oak can house five hundred species.

Under the ground, the tree's root system brings up nutrients from far below, keeps the soil structure open, works in symbiosis with fungi to break down organic matter and produce nutrients, protects from soil erosion and retains water in the soil. Apparently, the roots of different trees even form a kind of information system below the ground—a tree-Internet!

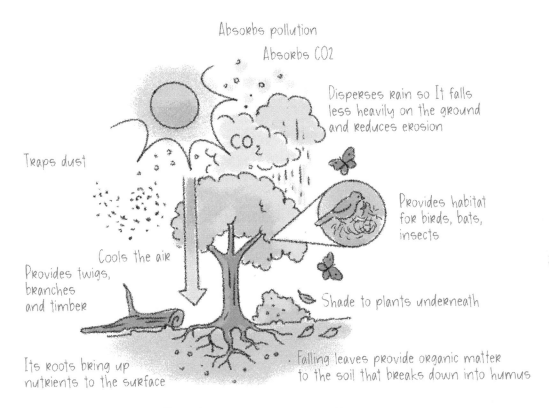

A TREE AND ITS FUNCTIONS

The falling leaves provide organic matter as a mulch that will break down into humus; the tree also provides twigs, branches and timber, which make for sustainable fuel or building material. Additionally, a tree can function as a windbreak and deliver fruit or nuts to you or to the birds, food for animals (which will leave manure to enrich the ground) and flowers for pollinating insects.

In permaculture, trees can be the basis of a *guild*, that is, related plants that work together for their mutual benefit. Particular trees have particular affinities with different insects, birds, animals, plants and may have particular functions; for instance, leguminous trees are nitrogen-fixing. So when we are choosing plants, we are choosing them to go together—we are choosing a family, a system.

MULTI-PURPOSE PLANTS

Other plants also have their functions. They have positive or negative relationships with other living things as well as with elements such as structures, soil, light, water and wind. Like trees, they trade phytochemicals and nutrients between their roots; for instance, goldenseal and native ginseng won't grow without each other. One protects the other from fungus; in return, it gets protection from animal attack.

Permaculture gardeners look for plants with as many functions as possible. This creates redundancy in the system and so makes your garden resilient; the plant is not doing a single job ('look nice' or 'fix nitrogen') but also others ('cover the ground to prevent weeds,' 'produce nutritious compost,' 'attract bees with its flowers,' 'deter pests with its smell').

Yes, you want to have tomatoes, strawberries, potatoes, some salad leaves, peppers, squashes. Maybe you actually like Brussels sprouts. But there are a few plants that you will find in most permaculture gardens; that's because in plant life, as in sport, there are specialists and there are all-rounders. And you need a few all-rounders on any team. They make all the other players better.

Specialist permaculture plants	
Plant – specialists	Functions
Comfrey *(symphytum officinale)*	All round garden helper Roots will break up the ground Bring nutrients up to the topsoil where other plants can get at them Big leaves distraction to slugs from other plants Flowers attract bees and hummingbirds Full of nitrogen and carbon - great mulch plant Medicinal - comfrey poultice can help aches and pains Perennial NOT edible
Borage *(borago officinalis)*	Good companion plant for strawberries, tomatoes and squash Good orchard plant Annual but self-seeding Concentrates trace minerals in the soil Protects against tomato hornworm and cabbage worms Bee attractor Calcium and potassium in leaves - good for compost Makes good infusion (ten minutes in boiling water)

Specialist permaculture plants	
Plant – specialists	Functions
Yarrow (*achillea millefolium*)	Compost activator Insect attractor (hoverflies, ladybugs) Natural fertilizer/green manure Can be used as a herb instead of tarragon Anti-inflammatory poultice Mildly toxic to cats and dogs
White Japanese rose (*Rosa rugos*)	Attracts insects, bees Ground cover or windbreak Petals edible in salads or candied Rose hips for tea or birds' winter food
Siberian pea shrub (*caragana arborescens*)	Nitrogen fixer Stabilizes soil with its roots Windbreak or hedgerow.
Oats (*avena sativa*)	Edible grain - porridge, bread, soup thickener Cover crop, over-winter catch crop Weed suppressant Increases nitrogen content Craft materials
Red clover (*trifolium pratense*)	Attracts useful predator insects Fixes nitrogen Improves soil texture and drainage Great in apple orchards A handful of the flowerettes added to buttered rice tastes like nectar
Elderberry (*sambucus nigra*)	Flowers and berries edible (make elderflower lemonade, berry jelly, syrup, dried berry tea) Medicinal - against colds and flu Good compost activator Leaves make 'tea' to spray - gets rid of aphids, carrot root fly, and peach tree borers
Anise hyssop (*agastache foeniculum*)	Bee plant Deters white fly and cabbage butterflies Pretty (purple flower spikes) Licorice flavor - herb for greasy/gamey meats Self-seeds

Specialist permaculture plants	
Plant – specialists	Functions
Alfalfa (medicago sativa)	Green manure crop. But more to it than that Nutrient accumulator - calcium, nitrogen, iron, phosphorus, potassium, magnesium - a really potent cocktail Roots go really deep and break up the soil even if it's hard clay - if you have problem soil this is your first go-to plant Insect attractor and compost activator You can eat alfalfa sprouting seed (note, use seed you've saved, not seed out of a seed packet which could be chemically treated), or use the leaves in salads and soups, even in smoothies
Burdock (arctium lappa)	Deep tap roots break up soil and bring nutrients to the surface Breaks up and reconditions soil Used in Japanese cooking (called 'gobo') Good detox herb - simmer 30 minutes for infusion
Daikon radish (Raphanus sativus var. longipinnatus)	Edible raw, pickled, baked, boiled, stir-fried Huge roots break up soil Nutrient accumulator Big leaves for mulch
Garlic (allium sativum)	Love it or hate it Edible (according to taste) Stimulates immune system, manages cholesterol and high blood pressure Repels aphids, codling moths (the nasties which make maggots inside apples), snails, carrot fly, and even deer, cabbage looper, peach tree borer and rabbits, flea beetles, spider mites Crushed garlic soaked overnight makes a spray against aphids, whiteflies, tomato and potato blight
Beetroot (beta vulgaris)	Big root opens up soil Nutrient accumulator Leaves for compost or mulch Edible boiled, roasted, pickled Anti-inflammatory, full of antioxidants
Other allium species	Mild onions and leeks right through to strong onion and garlic All parts of the plant are edible Increase other plants' resistance to disease and reduce insect infestation Juice of common onion can be used as moth repellent

Specialist permaculture plants	
Plant – specialists	Functions
Dandelion (*taraxacum officinale*)	Edible - leaves in salad Diuretic Tap roots up to a foot deep Nutrient accumulator Beneficial insect attractor Bee plant
American groundnut (*apios americana*)	Edible tubers Nitrogen fixer

These are your go-to plants for getting a lot of donkey work done, but most plants have several functions and will have good or bad relationships with other plants. So by knowing what functions a plant has, we know where to put it in relation to other plants. We may know that a plant does well in shade; if we also know that it's a nitrogen fixer, then it will be a good plant to put under a fruit tree that needs some help because it's not getting enough nitrogen.

There are some other good, multi-purpose plants, perhaps not quite as useful as those we've already mentioned, but worth knowing. The table below summarizes your options, including several trees and bushes.

MULCH MAKERS

One of my neighbors recently went shopping and came back with six expensive bags of bark chippings. They covered one rather small flower bed.

I didn't like to say anything, but you don't need to lay out money on bark chippings. You can make your own mulch! You can use straw, or cardboard, or wood chips or sawdust, but you can also do as the forest does and just use fallen leaves from trees and other plants—or even the whole plant, in some cases.

For instance, you could sow an entire patch with peas and beans. Underneath the soil, they're working to fix nitrogen for other plants to use. Then, when you're done with the crop, chop the legumes down and either leave them where they are or use them to mulch another area.

I like Jerusalem artichokes. I like them to eat; I like them because they're easy to grow and I also like them because they produce a huge amount of excellent mulch once the growing season is over. They're perennial, a really no-hassle crop.

'Weeds' like burdock, plantain, clover and mallow are plants many gardeners spend huge amounts of effort trying to eradicate. But why? Like comfrey, they make great mulch. And they are only 'weeds' because they're such great pioneers—they will grow anywhere they can get a foothold—so you don't need to pamper them. They'll just grow, then all you have to do is cut them down.

If you're worried that using weed mulch will let these weeds get a foothold where you don't want them, then you could use aquatic plants as mulch. Rush, sedge, reeds and cattails (*Typha latifolia*) make super mulches, but they can't grow in the garden as they need a pond to inhabit. If you're using a reed bed as part of your greywater harvesting system or already have a pond for fish or ducks, you can get it to do double duty by regularly harvesting a few armfuls of fresh mulch from it.

Plants don't need to be dead to be mulch! 'Ground cover' is just living mulch; for instance, a carpet of chamomile, clover, phacelia, or a blanket of mustard or sweet potatoes, will loosen up the soil with their roots, keep it shaded from the sun, provide nutrients and prevent weeds—that is, weeds you really *don't* want—getting established. You can use them as cover crops, too, when you don't have any vegetables growing in a particular bed (such as over winter); when you want to sow or plant edibles, just cut the ground cover plants down to use as green manure.

You can also use shrubs for green mulch. Nitrogen fixers like alders, locust bean trees and Russian, Siberian, or Pygmy pea shrubs grow quickly and can be trimmed or even cut right down to the ground to provide leafy branches for mulch.

If you want to make the minimum effort, don't grow your mulch plants on a separate area of the plot. Grow them with your crops—and when they're ready, 'chop and drop.' Just chop the plants off at soil level, leave the roots to die off in the ground, where they'll provide nutrients (even with perennials, some of the roots will die now they're no longer feeding the plant) and drop the rest of the plant on the ground. You might want to chop woody stalks or big leaves into smaller pieces just so they rot more quickly.

For even better results, use a sheet composting system; chop and drop, then add a layer of manure and top it off with another mulch. Instead of having a composting bin, the entire garden has effectively become your composting bin!

Good permaculture plants	
Plant	Functions
Lupine (*lupinus polyphyllus*)	Nitrogen fixer Bee plant Pretty flowers
Rhubarb (*rheum rhabarbarum*)	Food plant Leaves make good mulch Good cover plant
Lemon balm (*melissa officinalis*)	Makes a refreshing herbal tea, can be added to salads Bee plant, particularly later in the year Ground cover
Oregano (*origanum vulgare*)	Herb Ground cover Bee plant
Marigold (*calendula officinalis*)	Insect plant Petals make a good skin salve
Chamomile (*matricaria recutita*)	Makes relaxing herbal tea Good ground cover Self-seeds where ground is bare
Welsh (perennial) onion (*allium fistulosum*)	Anti-bacterial Anti-fungal (good for fruit trees) Edible
Horseradish (*armoracia rusticana*)	Helps prevent peach leaf curl Edible
Vervain (*verbena bonariensis*)	Butterfly plant Perennial Pretty pink-purple flowers
Black locust tree (*robinia pseudoacacia*)	Nitrogen fixer Insect attractor Windbreak Bee tree ** Allelopathic
Crab apple (*malus malus*)	Insect attractor Fruit can be used for jam, jelly Food for birds Pollinates regular apple trees Leaf mulch Sloe and wild cherry/plum/myrobalan have similar benefits and make a good edible hedge

Good permaculture plants	
Plant	**Functions**
Hazel (*Corylus avellana*)	Nuts Windbreak Can be coppiced to make sticks/canes Good animal habitat and forage
Paw-paw (*asimina triloba*)	Fruit Dye, fiber for crafts Leaf mulch
Mulberry (*morus alba*)	Berries great in puddings and pies Ornamental with pretty leaves Leaf mulch
Goumi (*elaeagnus multiflora*)	Nitrogen fixer Edible Hedgerow or windbreak Leaf mulch
Lavender (*lavandula*)	Used in cooking, fragrant, repels insects in closet Insect plant Windbreak
Sage (*salvia officinalis*)	Bee plant Herb
Chives (*allium schoenoprasum*)	Pest repellent Insectary Useful herb Nutrient accumulator
Creeping thyme (*thymus serpyllum*)	Edible herb Insectary Pest repellent Good ground cover for dryish soil
Jerusalem artichoke (*helianthus tuberosus*)	Edible (great soup) Insectary Big floppy leaves and tall stalks for mulch Windbreak Forage
Strawberry (*fragaria x ananassa*)	Nutrient accumulator Good ground cover Insectary Edible
Beans, peas	Nitrogen fixers Insect attractors Edible fresh or dried

COMFREY

BURDOCK

NUTRIENT ACCUMULATORS

These plants dig down deep in the soil to find particular nutrients and raise the nutrients through their stem and into their leaves. When the leaves fall, the nutrients are returned to the soil but stay in the top layer of soil where other plants can access them more easily. Well, that's the theory…

Unfortunately, a lot of what permaculture authors have had to say about nutrient accumulators is hearsay or just based on their own experience. And yes, I've added my own lot here, too. But there *has* been some real scientific research done on phytoaccumulation—plants' ability to concentrate minerals in their leaves—in connection with trying to salvage soils that have been contaminated with heavy minerals. The US Department of Agriculture, Dean Brown and Robert Kourik have been working to examine the science; further studies are being carried out at community farms.

But so far, the evidence is not conclusive. There is some strong evidence to support the fact that brake ferns are very efficient at drawing soil arsenic into their biomass, for example, but the claimed nutrient accumulation abilities of many other plants have not been proven scientifically. And you certainly shouldn't rely on it as your only way of providing degraded soil with more nutrients and better tilth. But then remember; one function, many elements? You shouldn't be relying on a single solution anyway. This is the great thing about permaculture; if one thing you're doing doesn't work, your other actions should take up the running instead.

In any case, most plants that are on the nutrient accumulators list also have big, strong taproots. So even if they're not lifting up all those minerals, they *are* breaking down your soil and improving its structure.

NITROGEN-FIXERS

These plants put nitrogen into the soil. Nitrogen is crucial for the growth of some plants—tomatoes and leafy greens, for instance, need plenty of nitrogen. But they can't get it out of the air; they have to find it in chemical forms such as ammonia or nitrate. Often, farmers and gardeners give them this in the form of chemical fertilizers.

However, you can use organic methods to provide the same resource by growing nitrogen-fixing plants. These plants work in partnership with microorganisms in their roots to turn atmospheric nitrogen into nitrogen-based fertilizers. Some of them partner with nitrogen-fixing bacteria of the *Frankia* genus; these plants include roses, oleasters (*Eleagnaceae*), buckthorns, bayberries and birch.

Others, mainly legumes (beans and peas), have different partner bacteria of the *Rhizobium* genus. These make the plant's roots form nodules to house the bacteria, which then begin their job of fixing nitrogen. In this family, you have alfalfa, beans, clover, cowpeas, lupines, vetches and soybean, as well as a plant you may not have thought of as a bean, the peanut!

There are nitrogen-fixing plants all the way up from tiny sprawling herbs to trees like black locust and Italian alder. Among shrubs, autumn olive (*Eleagnus umbellata*) and Sea buckthorn, as well as Russian olive (*Eleagnus angustifolia*, oleaster), can all help fix nitrogen for vegetables or fruit you're growing in their shade.

Traditional agricultural methods often included one nitrogen-fixing plant among the typical planting. For instance, the 'three sisters' corn-beans-squash system includes beans as a N-fixer. I wonder if the nursery rhyme "Oats and beans and barley grow" reflects a similar system?

Nitrogen fixing cover crops/ green manures	Nitrogen fixing shrubs	Nitrogen fixing trees	Nitrogen fixing flowers	Nitrogen fixing edible food/ vegetables	Nitrogen fixing herbs
Alfalfa (perennial)	Broom	Acacia	Bladder senna	Ahipa	Honeybush
Asparagus pea	Elaeagnus (various)	Alder	Californian lilac	Beans (bush)	Licorice (American)
Bean (fava/bell)	Gorse	Autumn olive	Chinese wisteria	Beans (garbanzo)	Licorice (European)
Bean (hyacinth)	Sea Buckthorn	Bayberry	Dyers greenweed	Beans (all others)	Rooibos
Bean (jack)	Siberian Pear	Black locust	Earthnut pea	Beans (snap)	
Bean (velvet)		California mountain mahogany	Glandular senna	Beans (string)	
Birds trefoil (perennial)		Cape broom	Indigo (all indigofera genus)	Breadroot (prairie turnip)	
Clover (arrowleaf)		Carob	Lupins	Chickpea	
Clover (balansa)		Cherry silverberry	Purple coral pea shrub	Jicama	
Clover (berseem)		Chinese yellow wood	Spring pea	Lentils	
Clover (crimson)		Chinese licorice	Tree lupin	Peanut (groundnut)	
Clover (mammoth red - perennial)		Evergreen laburnum	Wisteria (American)	Peas (green)	
Clover (New Zealand white - perennial)		Golden chain tree	Wisteria (Japanese)	Peas (snap)	
Clover (red - perennial)		Inga tree (tropical(	Wisteria (Kentucky)	Peas (snow)	
Clover (subterranean)		Japanese pagoda		Peas (sweet)	
Clover (sweet - perennial)		Kakabeak			
Clover (white - perennial)		Kentucky coffee bean			

Nitrogen fixing cover crops/ green manures	Nitrogen fixing shrubs	Nitrogen fixing trees
Comfrey Nitrogen fixer		Kowhai
Cowpea		Laburnum trees
Lespedeza (annual)		Locust tree
Lespedeza (sericea – perennial)		Mesquite trees
Medics		New jersey tea
Pea (field)		Persian silk tree
Pea (winter)		Purple coral pea shrub
Peanut (perennial)		Redbud/judas tree
Soybeans		Russian olive
Sun hemp		Seaberry
Velvet bean		Siberian pear
Vetch (bigflower)		Silverberry (eleagnus x ebbingei)
Vetch (chickling)		Silverthorn/ thorny olive
Vetch (common)		Silver wattle
Vetch (hairy)		Tagasaste
		Tamarind (tropical)

PEST REPELLENTS

In permaculture, as you'll have noticed, we often redefine a 'weed' as a 'pioneer plant.' But a pest is just a pest—something that's going to ruin your vegetables and fruit. These include slugs, snails, squash bugs and tomato hornworms, as well as butterflies whose caterpillars will eat your cabbages given half a chance. And aphids will eat nearly anything. Conventional gardeners use pesticides, or else they spend hours patrolling their gardens to pick up slugs and snails and throw them over the hedge into the neighbor's garden (which isn't really in line with the 'fair shares' ethic, is it?). Permaculturists use other things in the garden: ducks, chickens, garlic, ladybugs, all of which have their part to play in suppressing the pests.

You should have fewer problems with pests in a biodiverse garden, anyway. If you grow all your tomatoes in five straight rows with no other plants between, any pest that likes tomatoes knows exactly where to go. If each of your tomato vines is hidden between a lot of other, different plants, pests will have much more difficulty finding them.

Some essential oils repel bugs. Peppermint oil will get rid of squash bugs and ants (but also spiders); cedar and pine oils and hyssop deter slugs and snails. Rosemary oil can keep cabbage caterpillars at bay. If you grow these herbs and trees, you can also make an infusion to spray on the garden.

Growing the plants rather than making up a spray works too and unlike pesticides or spraying, it carries on working without your having to do anything. Dill is a prolific self-seeder and attractive plant, as well as an edible herb, but it will also discourage aphids, squash bugs and cabbage pests. Fennel will repel slugs and snails, as well as aphids. Peppermint is disliked by aphids, cabbage pests, squash bugs, winter flies and flea beetles.

It's often the scent that deters bugs; rosemary, lemon balm and catnip are all scented and pest-deterrent. And garlic will keep deer away.

If there's one plant in the garden that's the target for pests, it's tomatoes. So plant your tomato plants with cayenne, basil, garlic and onions; borage also repels tomato hornworm—so does tansy.

INSECTARY PLANTS

These plants don't repel pests; they attract beneficial insects. Remember that we need insects for pollinating your plants and that many insects are natural predators of pests. Ladybugs will eat aphids; they won't eat *all* the aphids, but they will certainly keep that population under control. In fact, it's often the larvae rather than the adults of the species who eat most of the pests—this goes for lacewings, too.

Ladybugs like fennel, dill and coriander. They also like yarrow, ajuga, Queen Anne's lace and the prairie sunflower (*Helianthus maximilianii*), as well as dandelions and hairy vetch (*Vica villosa*).

Yarrow, dill, dandelion, coriander, fennel and Queen Anne's lace will attract lacewings too, while they also enjoy caraway (not a well-known spice, but a very tasty one) and angelica. These plants (most of which are umbellifers, so have a number of short flower spikes spreading out like a parasol) also attract hoverflies, which can eat mealybugs as well as aphids, but hoverflies aren't choosy. They will also come to a whole load of other plants, including buckwheat, poached egg plant, sweet alyssum, penny-royal, spearmint, stone-crop and crimson thyme.

Parasitic mini-wasps control all kinds of moths, beetles and flies by laying their eggs inside the larvae of caterpillars. Again you'll find many of the same plants will attract these guys—yarrow, dill, caraway, coriander, for instance; they also like golden marguerite (*Anthemis tinctoria*), cosmos, butter and eggs (*Linaria vulgaris*), sweet alyssum, lemon balm (*Melissa*) and zinnia.

Golden marguerite, buckwheat, lemon balm, tansy and crimson thyme will attract tachinid flies which kill the caterpillars responsible for corn earworm, cabbage worms, stink bugs and a number of other pests.

Plants and flowers that repel pests	
Plant	Insects it repels
Allium	cabbage worms, aphids, slugs, carrot flies
Anise	snails, slugs, aphids
Asparagus	helps clear soil of root-knot nematodes attracted to tomatoes
Basil	flies, mosquitoes, carrot fly, white flies, asparagus beetles, tomato pests
Bay leaves	flies
Borage	tomato hornworm and cabbage worms
Catnip	mosquitoes, flies, deer ticks, cockroaches
Cayenne pepper	cabbage looper, spider mites, aphids, lace bugs, cabbage maggots
Chamomile	flies, mosquitoes, carrot fly
Chives	carrot fly, aphids, Japanese beetles, mites, snails & slugs
Chrysanthemums	ants, cockroaches, ticks, spider mites, fleas
Cilantro (coriander)	aphids, potato beetles and spider mites
Cosmos	corn earworm
Dill	aphids, squash bugs, spider mites, cabbage looper, small white
Fennel	aphids, slugs and snails
French marigold	tomato worm, slugs, nematodes, parasites
Garlic	root maggots, cabbage looper, peach tree borer and rabbits, aphids, flea beetles, spider mites, tomato pests, carrot fly, slugs
Geraniums	corn earworm and small white, leafhoppers
Horseradish	aphids, whiteflies, blister beetles, Colorado beetles, caterpillars
Hyssop	cabbage looper, small white, cabbage moths
Lavender	moths, fleas, mosquitoes, flies
Lemon balm	mosquitoes
Lemon thyme	mosquitoes
Lemongrass	mosquitos, flies, ticks, ants, gnats
Marigolds	aphids, mosquitoes, nematodes, whiteflies, asparagus beetle, tomato hornworm, cabbage maggot and moths
Mint	mosquitoes, cabbage moths, ants, rodents, flea beetles, fleas, aphids
Nasturtiums	whiteflies, squash bugs, aphids, bean beetles, cabbage loopers, cucumber eating pests, asparagus beetle
Onions	aphids, flea beetles, carrot flies, mosquitoes, Colorado potato beetles, whiteflies, bean beetles, cabbage worms, spider mites, moths
Oregano	many pests
Parsley	asparagus beetle
Pennyroyal	ants, fleas, ticks, chiggers, flies, gnats, mosquitoes

Plants and flowers that repel pests	
Plant	Insects it repels
Peppermint	aphids, cabbage looper, flea beetles, squash bugs, winter flies, small white
Petunias	aphids, tomato hornworms, asparagus beetles, leafhoppers, squash bugs
Pitcher plants	trap and ingest insects
Radish	cabbage maggot, cucumber beetles, squash bugs
Rosemary	mosquito, cabbage moths, flies, bean beetles, carrot flies, slugs, cabbage fly
Rue	aphids, fish moths, flea beetle, onion maggot, slugs, snails, flies, Japanese beetles
Sage	carrot fly, cabbage moths, snails, beetles, black flea beetles
Tansy	ants, beetles, flies, squash bugs, cutworms, whiteflies, tomato hornworm, small white, Japanese beetles, asparagus beetles
Thyme	cabbage moths, flies, corn earworm, whiteflies, tomato hornworms, small whites, cabbage fly, cabbage worm
Tobacco	carrot fly and flea beetles
Tomato	asparagus beetles

Pests that plants attract	
Plant	Pests it attracts
All or many	aphids, slugs, snails, sooty molds, swift moth caterpillar, tortrix moth caterpillar, wireworms
Agapanthus	agapanthus gall midge
Alder tree	alder leaf beetle, alder sucker
Alyssum	cyclamen mite, caterpillar
Apple tree	gall mites, apple leaf mining moth, apple sawfly, capsid bugs, citrus longhorn beetles, Asian longhorn beetles, codling moth, mussel scale, rosy apple aphid, small ermine moth, woolly aphid
Ash tree	lilac leaf-mining moth
Asparagus	asparagus beetle
Astrantia	astrantia leaf miner
Bay lauren	bay sucker
Bay trees	bay tree suckers, soft scale
Beans	bean seed fly, capsid bugs, eelworm
Beech tree	gall mites, beech red spider mite, beech scale, woolly beech aphid
Beetroot	beet leaf miner
Birch tree	gall mites, apple leaf mining moth, buff tip moth, citrus longhorn beetles, Asian longhorn beetles, vapourer moth

Pests that plants attract	
Plant	Pests it attracts
Blackberry	brown tail moth, blackberry aphid, raspberry beetle, red berry mite, scurfy rose scale, Japanese beetle, leafrollers, rednecked cane borer
Blackcurrant	blackcurrant big bud mite, blackcurrant gall midge
Box hedges	box sucker
Broad bean (seeds)	broad bean seed beetle, bean weevils
Broccoli	cabbage caterpillars, cabbage whitefly, flea beetles, cabbage worm, aphids
Brussel sprouts	aphids, leaf miners, cabbage loopers, slugs
Cabbage	cabbage caterpillars, cabbage root fly, cabbage whitefly, flea beetles, mealy cabbage aphid, cabbage worm, cutworm
Cauliflower	cabbage root fly, mealy cabbage aphid
Carrots	carrot fly, eelworm
Celeriac	carrot fly, celery leaf mining fly
Celery	slugs, carrot fly, celery leaf mining fly, nematodes
Cherries	cherry aphids, brown tail moth, cherry blackfly
Cherry tree	apple leaf mining moth, brown tail moth, pear, cherry slugworm
Chestnut tree	oriental chestnut gall wasp
Chilli	whiteflies, thrips, aphids, leaf miners, psyllids
Chives	allium leaf miner
Citrus trees	citrus longhorn beetles, Asian longhorn beetles, fluted scale, soft scale
Conifers	adelgids, conifer aphids
Corn	corn earworms, corn borer
Cucumbers	spider mite, glasshouse leafhopper, cabbage beetle
Currants	currant aphids, currant blister aphids (red, white, blackcurrants), gooseberry sawfly (red, white), currant scale
Dahlia	black bean aphid
Deciduous trees	lackey moth
Eggplant	glasshouse leafhopper, cutworms, tomato hornworms
Eucalyptus	eucalyptus gall wasp, eucalyptus sucker
Euonymus	euonymus scale
Evergreen euonymus	euonymus scale
French beans	black bean aphid
Fruit trees	spider mite, tawny mining bee, fruit aphid, spotted wing drosophila, winter moth caterpillar
Fruit plants	tawny mining bee, vine weevils, spotted wing drosophila
Garlic	allium leaf miner, leek moth, eelworm, nematodes

Pests that plants attract	
Plant	**Pests it attracts**
Gooseberry	*gooseberry sawfly, currant scale*
Greenhouse plants	*glasshouse whitefly, glasshouse leafhopper, glasshouse red spider mite, glasshouse thrips, hemispherical scale, mealybug, oleander scale*
Hawthorn tree	*apple leaf mining moth*
Hazel tree	*gall mites, citrus longhorn beetles, Asian longhorn beetles, vapourer moth*
Hedges	*woolly beech aphid*
Holm oak	*holm oak leaf-mining moths*
Horn beam tree	*buff tip moth*
Horse chestnut	*horse chestnut leaf-mining moth, horse chestnut scale*
Kohlrabi	*aphids, cabbage moth, cabbage white butterfly, flea beetle, slugs and snails*
Leeks	*allium leaf miner, leek moth*
Lettuce	*bean seed fly, slugs*
Lime tree	*buff tip moth, lime nail gall mite, oak slugworm, vapourer moth*
Melon	*aphids, cucumber beetle, squash beetle*
Mint	*brown tail moth, cabbage moth*
Nasturtium	*black bean aphid*
Nectarine	*currant scale*
Oak tree	*buff tip moth, oak gall wasp, oak processionary moth, oak slugworm, winter moth caterpillar*
Okra	*leaf hopper, red spider mite, root knot nematode, fruit borer, mealy bug*
Onions	*onion fly*
Orchids	*spider mite*
Oriental greens	*cabbage root fly*
Ornamental plants & vegetables	*slugs and snails, vine weevils*
Parsley	*carrot fly, celery leaf mining fly*
Parsnip	*carrot fly, celery leaf mining fly, eelworm, carrot weevil, leafhopper*
Peach	*currant scale*
Peanuts	*aphids, nematodes*
Pear tree	*pear leaf blister mite, codling moth, pear and cherry slugworm, pear blister mite, pear midge, pear-bedstraw aphid, social pear sawfly*
Peas	*slugs, pea weevils, pea moth, aphids, leaf miners*
Peppers	*glasshouse leafhopper*
Philadelphus	*black bean aphid*
Pine trees	*pine sawflies*

Pests that plants attract	
Plant	**Pests it attracts**
Plum tree	gall mites, plum fruit moth, brown tail moth, hawthorn webber moth, plum aphid
Potatoes	slugs, capsid bugs, potato cyst nematodes, beetles, aphids
Privet	privet aphid
Pumpkin	snails, slugs, beetles, squash bugs, vine borers, aphids
Radish	cabbage root fly, flea beetle, red spider mite
Raspberries	southern green shield bug, raspberry aphid, raspberry beetle, raspberry leaf, bud mite
Rocket salad	flea beetle, aphids, slugs
Rosemary	cuckoo spit, rosemary beetle, sage & ligurian leafhoppers, cabbage fly, bean beetle
Runner beans	aphids, slugs, snails, southern green shield bug
Sage	sage & ligurian leafhoppers, cabbage moth
Shallot	leek moth
Shrubs	lackey moth
Spinach	beet leaf miner
Spring onion	cutworms, onion maggots, slugs
Sprouts	cabbage caterpillars, cabbage root fly, cabbage whitefly, flea beetles, mealy cabbage aphid
Squash	black cucumber beetles, squash bugs, squash vine borer, aphids
Strawberries	spider mite, foliar nematodes, woodlice
Sunflowers	bean seed fly
Swede	cabbage root fly, mealy cabbage aphid
Sweet potato	wireworms, root-knot nematode, sweet potato whitefly
Swiss chard	beet leaf miner
Sycamore tree	gall mites, winter moth caterpillar
Thyme	cabbage fly, cabbage moth
Tomatoes	spider mite, glasshouse leafhopper, southern green shield bug, potato cyst nematodes, corn earworms, tomato hornworm
Turnip	cabbage root fly, flea beetles, cabbage aphids
Viburnum tinus	viburnum beetles
Walnut tree	gall mites, codling moth
Willow tree	large willow bark aphid
Zucchini	awl snails
Note – slugs and vine weevils eat many fruit & vegetables but the ones listed above are particularly affected	

As you see, a lot of these plants attract more than one species. However, don't just grow one; having a number of different plants doesn't just conform to the 'one function many elements' rule, but it helps ensure your beneficial insects have a number of different places to go, which might depend on where the sun is, or where the wind is coming from. Having a birdbath or some other water available also helps these insects—a little fountain surrounded by some of the herbs mentioned would be a powerful insect attractor.

Beneficial insects			
Insect	Type of benefit	Plants attracted to	Pests it controls
Assassin bug	Predator	Alfalfa, daisies, Queen Anne's lace, dandelions, goldenrod, tansy	Aphids, caterpillars, spider mites, insect eggs
Big eyed bug	Predator	Cosmos, fennel, goldenrod, mint, spearmint, alfalfa, caraway, sunflower	Aphids, caterpillars, mites, leafhoppers, spider mites
Braconid wasp	Pollinator, parasite	Flowering plants, carrots, catnip, chamomile, dill, fennel, feverfew, Queen Anne's lace, sweet alyssum	Aphids, beetles, hornworms, squash bugs, stink bugs, tent caterpillars
Butterfly	Pollinator	Sweet alyssum, asters, black-eyed Susans, butterfly weed, cardinal flowers, coneflowers, coreopsis, fennel, goldenrod, joe pye weed, milkweed, parsley, Queen Anne's lace, zinnias, chives, French marigold	n/a
Dragonfly and damsel fly	Predator	Backyard water features, creeks, ponds	Flies, gnats, mosquitos, small fish
Ground beetle	Predator	Clovers, amaranth plants, mulch, vetch	Slugs and snails, caterpillars, potato beetles, cutworms, cabbage worms, aphids
Hoverfly	Predator, pollinator	Oregano, garlic chives, sweet alyssum, buckwheat, bachelor buttons, fern-leaf yarrow, common yarrow, carpet bugleweed, lavender globe lily, basket of gold, dill, golden marguerite, dwarf alpine aster, masterwort, Queen Anne's lace, fennel, poached egg plant, lemon balm, tansy, spearmint, parsley, African marigold, cilantro, phacelia, pennyroyal, crimson thyme, chickory	Aphids, mealybugs and other garden pests

Beneficial insects			
Insect	Type of benefit	Plants attracted to	Pests it controls
Lacewing	Predator	Cilantro, cosmos, dill, fennel, ornamental onion, Queen Anne's lace, sunflowers, tansy, yarrow, golden marguerite, four-wing saltbush, purple poppy mallow, caraway, dandelion, angelica	Aphids, mealybugs, mites, thrips, white flies, insect eggs, caterpillars, cucumber beetle
Ladybird	Predator	Bergamot, cosmos, dill, goldenrod, lemon balm, mint, parsley, Queen Anne's lace, statice yarrow, zinnias, fern leaf yarrow, common yarrow, carpet bugleweed, basket of gold, golden marguerite, butterfly weed, four wing saltbush, CA buckwheat, fennel, prairie sunflower, cilantro, ajuga, dandelion, hairy vetch	Aphids, caterpillars, mealybugs, mites, scale, thrips
Minute pirate bug	Predator	Cosmos, caraway, alfalfa, spearmint, fennel, goldenrod	Mites, aphids, thrips, alfalfa weevil, armyworm, asparagus beetles, bean beetles, bean thrips, beet armyworm, beet leafhopper, brown almond mite, caterpillar eggs, coconut mealybug, grape leafhopper, greenhouse whitefly, leafhopper, mealybug, Mexican bean beetle, potato leafhopper, silverleaf whitefly, spider mite, sweet potato whitefly
Native bee	Pollinator	Basil, bergamot, blazing star, coneflowers, cucumber, dill, evening primrose, indigo, joe pye weed, lobelia, mint, oregano, peas, rosemary, sneezeweed, sweet alyssum, borage, chives, flowering trees, French marigold, phacelia, sunflower, comfrey, sage, creeping thyme, lavender, fennel, cilantro, lemon balm	n/a

Beneficial insects			
Insect	Type of benefit	Plants attracted to	Pests it controls
Parasitic wasp (ichneumonid wasp, chalcid wasp, tachinid fly. Also includes braconid wasp—see above)	Parasite	Cilantro, fennel, Queen Anne's lace, dill, caraway, yarrow, golden marguerite, cosmos, sweet alyssum, lemon balm, zinnia, flowering trees and shrubs, tansy, buckwheat, crimson thyme	Aphids, scale, whiteflies, sawfly larvae, ants, leaf miners, caterpillars, grubs, European corn borers, tomato hornworms, codling moths, cabbage loopers, cabbage worms, beetles
Rough stink bug	Predator	Trees, particularly fruit	Aphids, beetles, caterpillars
Soldier beetle	Predator, pollinator	Goldenrod, marigolds, zinnias	Caterpillars, aphids
Spined soldier bug	Predator	Catnip, goldenrod, hydrangea	Caterpillars, fly larvae, cabbage worms, potato beetles, grubs, aphids
Syrphidae	Pollinator, predator	Bergamot, cosmos, dill, goldenrod, lemon balm, mint, parsley, Queen Anne's lace, statice yarrow, zinnias	Aphids, caterpillars, mealybugs, scale, thrips
Tachinid fly	Predator	Queen Anne's lace, carrots, cilantro, dill, buckwheat, sweet clover, anise hyssop, golden marguerite, lemon balm, pennyroyal, parsley, tansy, crimson thyme	Grasshopper, beetles, bugs, caterpillars, earwigs
Trichogramma mini wasp	Predator	Fern leaf yarrow, common yarrow, lavender globe lily, dill, golden marguerite, masterwort, purple poppy mallow, caraway, cilantro, Queen Anne's lace, fennel, lemon balm, parsley, tansy, crimson thyme	Moths

WILDLIFE NURTURERS

Where some gardeners would put a bird feeder in a tree and think they've done their bit for wildlife, permaculture gardens usually also include plants that nurture wildlife. Permaculture looks at the garden holistically; it's much easier to include plants that have seeds or fruit that attract birds rather than heading out to the store to buy seeds and berries to put in a bird feeder.

Beneficial insects and plants attracted to									
Plant	Beneficial insect it attracts								
	Assassin bug	Big eyed bug	Braconid wasp	Butterfly	Dragonfly & damsel fly	Ground beetle	Hoverfly	Ladybird	Native bee
African marigold							•		
Alfalfa	•	•							
Amaranth plants						•			
Asters				•					
Bachelor buttons							•		
Backyard water features					•				
Basil									•
Basket of gold							•	•	
Bergamot								•	•
Black eyes susans				•					
Blazing star									•
Borage									•
Buckwheat							•	•	
Butterfly weed				•				•	
Camomile			•						
Caraway		•					•		
Cardinal flowers				•					
Carpet bugleweed							•	•	
Carrots			•						
Catnip			•						
Chamomile							•		
Chicory							•		•
Chives				•					•
Cilantro (coriander)							•	•	
Clover						•			
Common yarrow							•	•	
Coneflowers				•					•
Coreopsis				•					
Cosmos		•						•	
Creeks					•				

Beneficial insects and plants attracted to									
Plant	Beneficial insect it attracts								
	Assassin bug	Big eyed bug	Braconid wasp	Butterfly	Dragonfly & damsel fly	Ground beetle	Hoverfly	Ladybird	Native bee
Crimson thyme							•		
Cucumber									•
Cumin							•		•
Daisies	•								
Dandelion	•						•	•	
Dill			•				•	•	•
Dwarf alpine aster							•		
Evening primrose									•
Fennel		•	•	•			•	•	
Fern-leaf yarrow							•	•	
Feverfew			•				•		
Flowering plants			•						
Flowering trees									•
Four wing saltbush							•	•	
French marigold				•					•
Garlic							•		
Golden marguerite							•	•	
Goldenrod	•	•		•				•	
Hairy vetch								•	
Indigo									•
Joe pye weed				•					•
Lavender globe lily							•		
Lemon balm							•	•	
Lobelia									•
Masterwort							•		
Milkweed				•					
Mint		•						•	•
Mulch						•			

Beneficial insects and plants attracted to									
Plant	Beneficial insect it attracts								
	Assassin bug	Big eyed bug	Braconid wasp	Butterfly	Dragonfly & damsel fly	Ground beetle	Hoverfly	Ladybird	Native bee
Nasturtium				•			•		•
Oregano							•		•
Parsley			•				•	•	
Peas									•
Pennyroyal							•		
Phacelia							•		•
Poached egg plant							•		
Ponds					•				
Prairie sunflower								•	
Queen anne's lace	•		•	•			•	•	
Rosemary									•
Sneezeweed									•
Spearmint		•					•		
Statice yarrow								•	
Sunflowers		•						•	•
Sweet alyssum			•	•			•		•
Tansy	•							•	
Vetch						•			
Zinnias				•				•	

Beneficial insects and plants attracted to									
Plant	Beneficial insect it attracts								
	Parasitic wasp	Lacewing	Minute pirate bug	Rough stink bug	Soldier beetle	Spined soldier bug	Syrphidae	Tachinid fly	Trichogramma mini wasp
Alfalfa			•						
Angelica		•							
Anise hyssop								•	
Bergamot							•		
Buckwheat								•	
Caraway	•	•	•						•

Beneficial insects and plants attracted to									
Plant	Beneficial insect it attracts								
	Parasitic wasp	Lacewing	Minute pirate bug	Rough stink bug	Soldier beetle	Spined soldier bug	Syrphidae	Tachinid fly	Trichogramma mini wasp
Carrots								•	
Catnip						•			
Chicory	•								
Cilantro (coriander)	•	•						•	
Common yarrow	•								•
Cosmos	•	•	•				•		
Crimson thyme	•							•	•
Dandelion		•							
Dill	•	•					•	•	
Fennel	•	•	•						•
Fern-leaf yarrow	•	•							•
Flowering trees	•								
Four wing saltbush		•							
Fruit trees				•					
Golden marguerite	•	•						•	•
Goldenrod			•		•	•	•	•	
Hairy vetch	•		•						
Hydrangea						•			
Lavender globe lily	•								•
Lemon balm	•						•	•	•
Masterwort	•								•
Mint							•		
Ornamental onion		•							
Parsley	•						•	•	•
Pennyroyal	•							•	•
Purple poppy mallow		•							•
Queen anne's lace	•	•					•		•

Beneficial insects and plants attracted to									
Plant	Beneficial insect it attracts								
	Parasitic wasp	Lacewing	Minute pirate bug	Rough stink bug	Soldier beetle	Spined soldier bug	Syrphidae	Tachinid fly	Trichogramma mini wasp
Shrubs	•								
Spearmint			•						
Statice yarrow								•	
Sunflowers		•							
Sweet alyssum	•								•
Sweet clover								•	
Tansy	•	•						•	•
Zinnias	•				•		•		

All kinds of shrubs with small fruit are loved by birds; elderberry, blueberry, roses (which have rose hips in autumn), ceanothus, wild cherries, myrobalan, hawthorn, dogwood and chokeberry. An additional benefit is that the birds will drop guano under the shrub, which helps fertilize the soil.

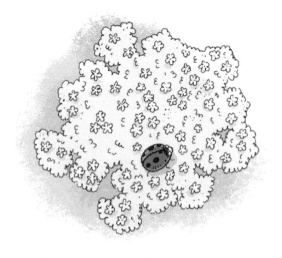

LADYBIRD ON QUEEN ANNE'S LACE

FORTRESS PLANTS

Every gardener has a story to tell of invasive plants. When you've spent a lot of time on your garden, it's heartbreaking when some bamboo from next door decides to colonize your space or when your mint patch takes over all the other herbs.

Fortress plants are your first line of defense against invasion. They produce a huge wall—above ground, thick branches and foliage; below ground, tough roots. So they can stop couch grass, creeping vines and other weeds that spread by root or by suckering, as well as self-seeders you don't want. Some of them are mildly toxic to other plants as well.

Comfrey, the top multi-function plant, is a great fortress. So are Jerusalem artichokes, red hot poker and lemongrass.

WINDBREAKERS

Associated in function with the fortress plants, windbreakers are plants that will make good hedges to break the force of the wind, provide privacy, create sun-traps and even possibly keep out browsing deer.

PLANT A HEDGE TO KEEP DEER OUT

Edible hedge plants can form an effective barrier; you might alternate hazel, crab apple, sloes and wild cherries with spikier plants like hawthorn and gooseberries. Sea buckthorn is spiky and produces very nutritious orange fruits.

On the other hand, if you're just after a windbreak, Jerusalem artichokes do well, standing tall and close and casting a green shade with their floppy leaves. You could use willow, which will make a living fence and can, if you trim it regularly, provide materials for making fences and baskets; or you could plant Maximilian sunflowers or bamboo (which, again, provides materials for basketry and other crafts, as well as garden canes).

ROOT RESTORERS

Plants with spiky taproots that go drilling deep in the soil are fantastic for reclaiming compacted and clay-rich soil. 'Spike roots' can be sown for the first couple of years before you make your final planting to do their magic and make the soil softer. Or you can sow them between other plants to make sure the soil stays that way. Additionally, smaller parts of the roots decay during the winter or after you have cut the top plant, adding organic material to the soil.

Daikon, dandelion, mustard, alfalfa, rapeseed (colza) and chicory are all great root workers. And guess what? Comfrey is on this list too!

ANNUALS VERSUS PERENNIALS

Annual plants have to be planted fresh from seed every year; they don't survive the winter. Perennial plants, even if their leaves die down, remain in place and most will flourish again next year; however it may be worth noting that some perennials are frost-tender. You can probably guess which of these takes more work, so if you have a choice, plant a perennial!

Despite this, eighty percent of the food we eat comes from annual plants. Grain, tomatoes, squash, corn, most salad leaves, rice—all of this comes from annuals. But using permaculture, you can switch things around a bit (remember, the bit of our diet that comes from perennials already is probably the part you most enjoy eating—strawberries, apples, peaches and other fruit and nuts)!

There are other reasons why we ought to grow more perennials. Unlike annuals, we don't need to disturb the soil to plant or sow afresh every year; we don't compact the soil by walking on it and it's not likely to be eroded if we don't crumble it.

Perennials have deeper roots. Annuals don't invest much in their root system; what's important to them is getting to set seeds and reproduce before the end of the season. So they have shallow roots, which don't help to stabilize the soil and don't reach nutrients or water much more than six inches below the surface. Then they leave the soil bare in winter, so it can be eroded, or weeds can move in.

Perennials grow more slowly, but they make a big investment in their roots. That lets them reach extra nutrients; it also helps stabilize the soil. Tree roots can run up to sixty meters deep—as deep as the tree is high. That means they recycle soil nutrients instead of just taking them out as annuals do. They also get started more quickly in the spring since they are just growing a bit more—they don't have to start from scratch.

You have more standing biomass with perennials—that is, more energy stored in the plant. Look at a tree and you can see how energy is stored in the wood, in the twigs, in the leaves; it's also stored in the roots, though you can't see those. That's all available for you to use—leaves and twigs for mulch and wood for timber, eventually.

The only difficulty is that in temperate climates, there aren't so many choices of perennial. You have more in tropical climates. Sometimes, annual vegetables do have a perennial version, like Nine Star perennial broccoli, but you may have to look through a lot of catalogs to find them or exchange with another permaculture gardener. Sweet potato is perennial and easy to grow and the leaves can be cooked (but *not* eaten raw).

Besides fruit trees and berries, you can plant other perennials: asparagus, all the mints, rhubarb, taro (in zones 7-10), sorrel, lovage, chard, kale and Turkish rocket. Also, Jerusalem and globe artichokes, salsify, and Welsh onions, for example.

While they're not quite perennial, you can also save yourself time by choosing annuals that self-seed pervasively. Many herbs, like cilantro, dill and fennel, will reseed; so will orach (mountain spinach), mustard and nasturtiums.

Although so far there are no varieties available for the garden, the Land Institute in Kansas—as well as some other institutions—has been working on turning annual varieties of grains and pulses into perennials. Ancient varieties of grain were perennials, but through centuries of breeding, they became annual. Now, maybe, we can reverse the process.

SHRUBS

I've seen plenty of gardens with a lovely orchard and a productive vegetable garden, but nothing else in between. But shrubs are at the heart of any permaculture garden; even if you don't have room for a tree, like on a balcony, you have room for a couple of small shrubs, which will provide a canopy layer for other, smaller plants.

In the garden, shrubs have several functions to fulfill. They can be a windbreak at a low level, which stops trees, creating a wind tunnel effect. You can use them to create a hedge, a diverse habitat that will shelter wildlife as well as smaller plants and which you can use to create a shady microclimate if your garden gets sun that's too strong for some of your more delicate plants.

They're perennials, so they don't need much more than mulching and pruning once they've got started. Some are nitrogen fixers; their roots will stabilize earth—for instance, along a bank—and best of all, you can pick shrubs that will feed you.

So blackberries, raspberries and their various relatives—goji, gooseberries, red, white and blackcurrants—hazel and crab apple, all have a place in your garden. If you're in one of the colder areas of the world, honeyberries are remarkably cold-tolerant (down to minus forty-eight centigrade!), while blueberries, depending on variety, have you covered all the way from zone 4 to zone 10.

By the way, don't forget vines such as the grape and kiwi fruit. They're particularly useful for stacking plants vertically, as you can train them up a trellis or just let them wind around a large tree. They will need pruning, though, or they'll take over—and with more stems and leaves, there will be less energy for fruit (put your grapes in zone 1 so you are close to them and can easily do a little work every so often). Grapes are more productive when trained using a formal method, such as the Guyot system—there are many guides for this available online.

HEDGES

Growing an edible hedge fulfills many functions in your garden. It gives you food, it can enhance your privacy, act as a windbreak, stabilize soil and increase biodiversity in the garden. A hedge also makes a great riparian buffer, protecting a stream or pond from runoff; it can deter livestock—whether wandering deer or your own sheep—from getting into the vegetable bed.

Not every plant in the edible hedge needs to be edible. But you can start with hazelnuts and wild or bush varieties of fruit like crab apple, cherry and plum; you can add bushes like redcurrant or blackcurrant and climbing plants like blackberries. Sloes (blackthorn) are also worth growing and can be used to make sloe gin or jelly, plus the thorns on the bush will stop livestock pushing through, while birds can still get the fruit. Growing roses in the hedge gives you rose hips as well as flowers in season. Elder gives flowers (which can be fried or used to make cordial or wine) and berries for jam or jelly or pies.

Think about what you grow in the under-story of the hedge, too, because every hedge becomes its own little micro-environment. Mosses and lichens do well, but so do many edible annuals.

FRUIT TREES

If shrubs are high yielding and low maintenance, fruit trees are even better! Many will live over a century, though they do take some time to get established. Apples can take three or four years to produce a proper crop, while stone fruit will usually start producing in year four.

However, if you are buying a tree on a full-size vigorous rootstock like M25, the tree will spend a lot of its time in the early years just growing, so you might have to wait a while for fruit. On the other hand, on rootstocks like Bud.9 or M9, M26, M27, which are intended to give smaller trees, you can get a crop a little earlier, though the tree won't live as long.

The best time to plant bare-root trees is in the fall, while they are dormant. If you can't plant them the same day, then 'heel them in'—dig a trench with a sloping side, water it well, put all the trees in it and then cover up the roots. You have to keep the roots from drying out. If you've dug a swale, you can prepare the tree planting holes on the berm before you collect your trees if you organize with the nursery. You could also buy mail order, but if you do, get your order in early, before the end of August. I prefer to buy from a nursery because they know what works in our particular climate.

When you plant the tree, water the hole the day before, so the ground is well hydrated. Give its roots room to grow by loosening the soil at the sides and bottom of the hole with a fork, so the roots have some openings they can take. At the same time, put the tree roots in a big bucket so they can drink in moisture.

When it's time to plant, create a small mound of good soil in the middle of the hole, put the tree on top of this and then spread the roots around so they form a rough cone. Then cover over the roots; if you have high winds where you are, stake the tree firmly.

Remember to check that a tree is in the right place. Figs in many northern climates will bear fruit, but only if they're in a sunny corner with a wall to reflect the sunlight on them. They also do better with restricted roots and used to be grown in a suitcase in Victorian times. Apples, cherries and pears are really robust; damsons, plums, greengages, peaches and apricots need more sun and, in particular, are vulnerable to late frosts. On the other hand, in zones 5-9, you can grow mulberries, peaches, nectarines; citrus will only grow outside if you're in zone 9 or warmer (but can be grown in pots and overwintered indoors). Papaya (actually a shrub) will grow in zone 10, as will banana and its cousin, plantain. American persimmons or kaki are also very tasty.

You also need to check whether a tree is self-fertile or whether it needs another, different variety tree to pollinate it. For instance, French gardeners plant cherries in pairs—usually Coeur de Pigeon and Napoleon, or Napoleon and Burlat (sour cherries, like Morello, on the other hand, are auto-fertile). Don't rely on there being the right tree somewhere around the place; make sure you have all the pollinators that you need. Make sure that they bloom at the same time, too!

For small gardens, some nurseries produce trees with three pollinators grafted onto the same rootstock. That can really help if you don't have room for three apple trees or two plum trees.

You might also think about staggering your crops by planting trees with different dates of bloom and harvest. With apples, you could get your early apples in September, mid-season in October and late-keepers all the way through to January; and the keepers can keep right through to May (earlies, on the other hand, need eating up fast).

Finally, the best way to plant your fruit trees is in a guild—an association of plants, all of which help each other. A guild is a small, self-sustaining system within your garden. We'll talk about this later, but remember, when you're choosing plants, you don't choose a single plant, you choose a whole family of plants to live and work together.

OTHER TREES

If you have a larger garden or a farm, you can start thinking about trees that don't produce fruit. For instance, oak is a marvelous tree; it yields up acorns for animal fodder and bird food, makes leaf mold for mulch and comes in different varieties which can live in different climate zones, from swamps to dry areas. It will grow huge and it will become its own environmental zone, harboring wildlife from nuthatches and woodpeckers to squirrels and raccoons. Sheep, goats and other animals will nibble on the foliage and young branches.

Young oak branches deliver splints for basketry and acorns can be dried, ground down and made into ash cakes on an open fire. Oaks were used to build ships of the Royal Navy in Britain and cabinetmakers love oak for its lively figure. It's a good over-story, canopy tree—but not for the suburbs.

Nut trees include the walnut, a large tree and one which is not particularly sociable as it's allelopathic—its leaves and roots contain a chemical that inhibits growth in other plants. Pecan, almond and macadamia are options for milder climates.

ROOTS AND TUBERS

Roots and tubers make good eating—as you know, the humble potato is a great vegetable. It can make roast potatoes, mashed potatoes, French fries, tartiflette, hash browns… but it's by no means the only tuber that's edible. Taro, sweet potatoes, Jerusalem artichokes and American groundnut (*Apios americana*) are all perennials—just harvest as many tubers as you need and leave the rest to grow. Chinese artichokes (*Stachys affinis*—actually a kind of mint) have knobbly little tubers, which you can dig up once the frost has killed off the top. These plants may be better planted in a guild or in their own area, rather than in the vegetable bed, as they can spread quite fast—that's what they were designed to do!

And there's one rather special tuber if you have a pond. Wapato or arrowhead needs to be grown in a pond or swamp, but it's a great tuber, tasting a bit like yam or sweet potato.

EDIBLE 'WEEDS'

The twenty-first-century human is a remarkably picky eater. We eat such a small percentage of the plants that could form part of our diet. And some of them we call 'weeds,' and we spend vast amounts of time trying to eradicate them.

Weeds are pioneer plants and they're often the first to sprout in spring, trying to steal a march on your nice salad leaves. Chenopodium album or lamb's quarter is supposed to be a weed and it will spring up all over—it tastes fine. Chickweed and corn spurry (*Spergula arvensis*) are other early-leafing weeds that can be chopped up into a salad, perhaps with a few seedlings you've weeded out from your radish salad or beet sowings.

Dandelions are edible; the leaves are quite bitter, so you wouldn't eat them on their own, but you can use them to give a bit of bite to a salad with other, less pungent leaves. They are best eaten when young and pale and in small quantities due to them being diuretic! The flowers also make great homemade wine. Sow thistle is full of antioxidants and again makes a great salad leaf. And purslane, which is found just about everywhere in the world, is another leaf with benefits; it has loads of iron. Plus, we have the ubiquitous stinging nettle. You *don't* want to put that in a salad, but it makes soup; rather boring soup, I always think but look, that's what all your other herbs and your garlic and onions are for; making things tasty! Failing that, it makes a great nitrogenous tea fertilizer.

The great thing with the weeds is that they'll grow anyway, they'll cover up bare soil and once you have other plants established, you can chop the weeds up and make manure. Plus… one or two of them you'll probably really enjoy eating.

CREATING MICROCLIMATES IN YOUR GARDEN

Each plant has its preferred conditions; it needs a certain amount of sun, a certain amount of moisture, the right soil type, soil acidity, place in the ecosystem. Very often, you will hear definitive statements like "this plant can tolerate zones 4-7, needs full sun, will not survive frost". In fact, that may or may not be true.

The reason why it's not *necessarily* true is that you can create microclimates in your garden if you're smart enough. An easy example is if you create a greywater harvesting system that includes a reed bed; suddenly, you have a wetland environment. That might let you grow some plants for which your soil would otherwise be too dry. Or, if you paint a south-facing wall white and shelter it from wind, mulching the roots really well, you'll have a microclimate that is a couple of zones warmer than the rest of the garden—this is the way gardeners in cooler climates manage to grow figs, apricots and peaches.

Another way to make a sun trap is to plant trees, then shrubs, in an open 'U' facing south. You can use slopes, too; a south-facing slope will get more sun than a north-facing slope; south-east will get gentle morning sun, south-west will get hot, afternoon sun. Every wine grower anywhere in the world knows that!

With forest gardening and guilds, you also build 'edge' microclimates—plants that thrive in your apple tree's shade might not do so well at the drip line, but other plants which need more sun will love it. The drip-line is the furthest reach of a tree's branches. If rain falls on the tree, the raindrops will run all the way down the branches to the last twig, and where they fall is the drip-line.

Be attentive to the microclimates in your garden and your plants will love you. And they'll show it, with lots of flowers, fruit, vegetables and nuts.

CREATING A MICROCLIMATE

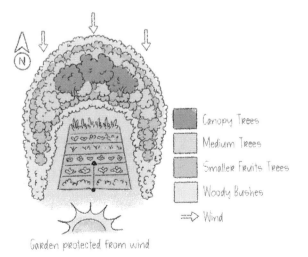

CREATING A SUNTRAP MICROCLIMATE

EXERCISES

Adventures in the kitchen

How many plants do you regularly eat? Do you know where they grow?

You may be amazed at how few different types of plant you eat. What haven't you tried? Make a list and try them.

Which vegetables have you never tried raw or pickled? Try those, too. Japanese, Korean and Middle Eastern cuisines, in particular, have some excellent pickles.

You may feel a bit differently about growing them in your garden now!

Heritage varieties

Many modern vegetables have been developed to be grown in monocultures to maximize yield, not taste. Look up some heritage varieties of tomato, apple and other edible plants.

Why did they fall out of fashion?

Do you think some would be worth growing? In some countries, there are heritage seed libraries that collect such seeds and look for willing gardeners to propagate them.

You might also like to look up varieties that used to be grown a lot in your area. Many places have their own specifically adapted varieties and the history can be quite interesting.

Cultivate Your Paradise: A Review Request for "Creating Your Permaculture Heaven"

Unlock the Power of Green Generosity

"Planting seeds of knowledge grows a garden of possibilities." - Nydia Needham

Hey Future Permaculture Pros,

Did you know that permaculture isn't just a modern idea? In the Himalayas and Kerala, gardens thrive using permaculture principles – combining trees, shrubs, and salad plants in a dance of biodiversity. These ancient techniques enriched the soil, creating a black and rich haven for abundant crops. The magic lies in understanding nature, using polycultures where different plants share space, just like nature intended.

Why Permaculture Matters: Permaculture is like a superhero for your garden. It's not just a set of rules; it's a design approach that mimics natural systems, making your garden dynamic, alive, and full of possibilities. Whether you see it as a philosophy, a toolbox, or a methodology, the key is not to forget the basic principles.

Your Mission, Should You Choose to Accept: I want to make "Creating Your Permaculture Heaven" accessible to everyone, especially those dreaming of turning their garden into a vibrant, sustainable paradise. But, here's where you come in – with your help, we can reach every aspiring permaculturist out there. Your review is the key.

How You Can Help:

1. Scan the QR code below to leave your review on Amazon.
2. Share your thoughts and let others know how this book can guide them to create their permaculture haven.

Your Review Could Help:

- Another small permaculture business flourish, providing for their community.
- An aspiring entrepreneur support their family with a sustainable garden.
- A fellow permaculturist find meaningful work in cultivating a permaculture paradise.
- A reader transform their garden, contributing to a more sustainable planet.
- One more dream of a green, permaculture-inspired world come true.

Join the Green Revolution: Leaving a review takes less than 60 seconds, costs nothing, and could change a fellow permaculturist's life forever. Scan the QR code below to leave your review (note this is the link for Amazon US, if you live in a different country, simply change the .com to the one for your country):

If you feel good about helping a faceless permaculturist, you're my kind of person. Welcome to our green club. You're one of us.

I'm super excited to share more tips and strategies in the coming chapters that will help you achieve your permaculture dreams faster and easier than you can imagine.

Thank you from the roots of my heart,

Nydia

PS - Fun Fact: When you share something valuable, you become more valuable too. If you believe "Creating Your Permaculture Heaven" can help another permaculturist, send this book their way — let's grow this green revolution together.

8

ATTRACTING BIRDS, BEES AND INSECTS TO YOUR GARDEN

Permaculture considers the whole ecology of the garden, not just plants. Insects pollinate your flowers, birds leave guano that fertilizes the soil, bees—if you do a little extra work—will make honey for you. The right animals can help, too: lizards can eat pests, frogs will eat slugs and snails. Your job, when you're designing a permaculture garden, is to ensure you have the right plants to attract them.

You might also add your own small livestock—chickens, ducks, rabbits, or even possibly goats. Again, they're part of your overall ecosystem; they have needs (in the case of chickens, grit is important, as it's how they grind their food down since they don't have teeth) and they have yields (eggs, guano, manure, milk—and in the case of goats, great entertainment value).

European readers can also attract hedgehogs (which aren't native to America), which are both cute and useful. They love eating slugs and snails (a *very* good reason for not using pesticides, which kill your hedgehogs, then your slugs will come back anyway). Make some piles of leaves towards the beginning of autumn and a hog may decide to hibernate there, though they wander very widely, so there's no guarantee he'll stay unless you have the slugs for him (by the way, if you ever make a bonfire, check for hibernating hogs and other animals before you light it).

In the wrong place, animals can be destructive. Deer let into a garden will eat anything. So will goats. Rabbits can eat the bark of young trees and kill the tree, as well as munching through leafy veg. But in the right place, or with the right persuasion, animals can help your permaculture garden prosper.

They do so many things. Earthworms till the soil for you, without exposing the soil to erosion or compacting it, or activating weed seeds, as mechanical tillage does. Bees and other insects pollinate. Birds sometimes spread seed—mistletoe, for instance, has seeds that are eaten by birds and grow once the bird has excreted, or in some cases regurgitated, on another tree. Birds and animals can dispose of waste for you and circulate nutrients in the garden. You can also eat many birds and animals (if you are not vegetarian).

But you need to keep the system in balance. You're effectively setting in place an ecosystem that will respond to little nudges every so often or where you control a key element (for instance, chicken tractors, which we'll talk about later). For example, with chickens, when you eat eggs, you're stopping the birds from having too many chicks, so you're helping to keep the chicken population in balance with the needs of your garden.

The more niches you can create, the more wildlife you'll get. If you have a pond, amphibians are happy, birds will bathe, insects will multiply and land animals can drink from it. Piles of dead logs, rocks, stone walls, thick bracken, and even the compost heap can provide niches for particular forms of wildlife. Keep your garden biodiverse!

My friend has cats. They do catch mice. They don't catch birds, but maybe they discourage birds from coming into some parts of the garden. They also limit the growth of her catnip herb. And they don't produce anything… except a lot of affection and purring.

Where do other animals fit? Apart from pets and micro-fauna, not in zone 0 (the house) or zone 1 (the intensively cultivated part of the garden, unless you want to keep guinea pigs or chickens, which can be kept here). But in zone 2, you might have fruit trees and bushes and chickens—who will eat some of your food scraps and keep the undergrowth free of slugs and bad bugs. In zone 3, you might keep larger or noisier animals like geese, goats, sheep, or bees, although some people may choose to keep bees in zone 2.

Your zone 4 of 'minimal maintenance' and zone 5 wilderness zone are where you might have low-maintenance livestock: deer, pigs, cattle. In zone 5, you'll probably have native fauna of all kinds.

But you might also think about succession. Just as plants have succession, with pioneer 'weeds' first, then shrubs and eventually full forest, you can use an animal succession if you're starting a garden. So you could start with a cow, or goats, or sheep (if you want mini-sheep, the French Ouessant breed is charming, standing less than nineteen inches high and producing a thick fleece; a few farms in the US now farm three-quarters Ouessant stock). You may be able to borrow livestock from a local small farmer. If you like exotica and have a large piece of land, alpacas and llamas are hardy, produce wool and are good pack animals. This livestock will clear up a lot of the vegetation and manure the ground; move them on, then introduce your poultry to finish the job.

Rotational grazing just keeps this succession going; the big guys clear the field, then the chickens come in and pick over what's left, and you just keep going. There's an extra benefit, as the chickens eat up parasitic larvae, leaving the field clean of these nasties that would otherwise be a health risk to your livestock next time they come to this field.

WORMS!

Earthworms and small insects help drag nutrients from mulch or dead vegetation down into the soil. They help to break it down so that it can decompose faster. Worm castings are full of nutrients, so pick them up and use them as compost. Earthworms also create channels in the soil, helping air and water to penetrate. Water drains ten times faster when the worms have been at work.

This is why we can do no-dig gardening because the earthworms do the digging—and without compacting the soil! You may not realize what a powerful force worms are. The average hectare of land contains seven million worms and they can turn over fifty metric tons of soil a year. That's the same weight as an adult whale. Even bugs are useful. Some bugs pollinate your plants; others prey on pests.

'BIG AGRICULTURE' VS. 'BUG AGRICULTURE'

Monoculture takes away insect habitats. Pesticides get rid of 'bad bugs,' but they get rid of beneficial bugs too. The problem with that is that often, the bad bugs recover and breed faster than the good ones. So then you end up with a worse problem than you started with;

WORM DIGGING INTO SOIL

instead of bad bugs with natural predators, you have bad bugs with no predators at all.

'Bug agriculture' encourages predator insects. Some of them simply eat their prey; others are parasitic, laying eggs inside other insects or their larvae. Some of them are general predators, others have a very specific prey, and the latter are particularly good to have, as they won't harm any of your helpful insects.

There's a whole family of parasitic wasps—braconid wasps, ichneumonid wasps, chalcid wasps and tachinid flies. If you ever see a caterpillar with what looks like little white balls on its back, those are the eggs of a parasitic wasp. Other types of wasp eat aphids, basically drying up the aphid from the inside till it's like an Egyptian mummy and then bursting their way out of it like the thing from 'Alien.'

By the way, although they're called wasps, most of these parasitic wasps don't look anything like the kind that stings. Chalcid wasps, for example, are tiny and a beautiful metallic green or blue.

The adult wasps feed on pollen and nectar, so flowers are critical for their survival. If you see a permaculture garden with no flowers, something's going wrong!

Hoverflies are also aphid-eaters and they fulfill the function of pollinators, too. You need bees, hoverflies and similar insects around the garden; otherwise, you would have to either rely on the wind or go round the garden

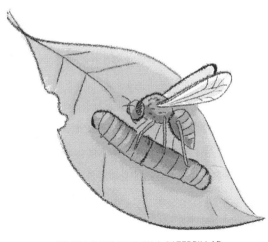
WASP LAYING EGGS ON A CATERPILLAR

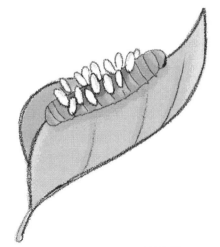
WASP EGGS ON A CATERPILLAR

pollinating every single flower yourself. If you look at an apple tree in bloom, you can imagine how hard that would be.

Ladybugs are cute but effective. Both the adults and the larvae feed on pests, particularly aphids, mites and mealybugs and their eggs. They are good at warning off predators—their bright red bodies are a telltale sign, but they also emit a nasty-tasting liquid—so once they're installed in your garden, as long as they find enough pests to eat, they're going to stay around.

You may not see your ground beetles, as they're mainly nocturnal; during the day, they scuttle underneath rocks or logs. At night, they'll eat caterpillars, potato beetles, cutworms, snails and slugs. Make sure to provide them with good habitat and they'll reward you with some hard pest-control work.

Lacewing larvae have huge (for their size) jaws, enabling them to grab and immobilize their prey. They can each guzzle up to forty aphids a day and have a huge appetite for mealybugs, insect eggs and caterpillars too. Once the larva becomes a lacewing, it's one of the most attractive insects in your garden, with delicately green-veined wings.

Finally, let's introduce a truly minuscule helper, the minute pirate bug—just two millimeters long. It punches above its weight, as it will take on a caterpillar many times its size if it's hungry. They're useful to have in your greenhouse, as they eat mites, thrips and aphids.

Some bugs also eat particular plants that you don't want in the garden. So weed-feeders can be quite useful. Cornell University has done work on these agents of biological control and it mentions several species. Purple loosestrife, a native plant in the UK, is controlled there by large insect populations; it has become an endemic weed in the US because those insects don't live there. Introducing the insects should help control the plant.

ATTRACTING BENEFICIAL INSECTS

Let's just talk about the different categories of beneficial insects before we start thinking about how to attract them. There are:

- Predators, which attack and eat the target pests.
- Parasites, which lay eggs on the target pests—the larva/caterpillar then eats the pest.
- Pollinators, which help to pollinate your fruit and vegetable plants.
- Weed-feeders, which eat plants you don't want.

Insects need food, shelter and water. A small pond can be useful in keeping your beneficial insect population happy. Some, like ground beetles, want a log-pile to shelter in, or loose bark on trees that they can wriggle behind, or mulch they can hide underneath. They also need the right conditions to reproduce.

And they need prey… which is why you won't use pesticides and you will need to tolerate a certain level of pests. The aim of a permaculture garden isn't zero pests but keeping the pests from doing excessive damage, naturally.

Most also need flowers. These may not be the same flowers conventional gardeners like (dahlias, double roses, annual geraniums), but there should be flowers—purple spires of sage, bright blue borage flowers among their prickly leaves, maybe lavender or tiny creeping thyme. The best flowers for pollination are simple flowers; the more a flower has been bred to have extra petals, like 'double' roses and pinks, the less attractive it is to pollinators.

But though we know that some plants are particularly attractive to insects and bees, your job as the designer of a permaculture garden isn't really about choosing the 'best' plants. It's about choosing a selection of plants that will start flowering as early in the spring as possible and carry on all year, right up to the onset of winter, so insects always have a reason to come to your garden. If you have a fantastic spring display but, like some gardens, no flowers at all after the end of June, you're wasting the opportunity to get the pollinators working for you on other things that need pollinating, like your courgettes/zucchini! And diversity is also important, as some species will prefer one or another plant.

Also, some insects like one kind of plant when they're adults, but the larvae prefer something quite different. So again, diversity is more important than just the 'right' flowers. A hedgerow with flowering shrubs and different smaller flowers beneath is a good way to provide a diverse ecosystem that will appeal to insects.

Particularly good plants for insects include:

- Zinnia
- Sunflowers
- Thyme, rosemary, mint
- Umbellifers like Queen Anne's Lace, fennel, dill, angelica, coriander
- Clover
- Chamomile and daisies
- Dandelions—one of the earliest flowers to come out, so it helps fill a gap
- Marigolds, cosmos, lemon balm, phacelia

An insect hotel can provide homes for insects that usually nest in deadwood. That's a minority—about twenty-five percent of bees, for instance—but worth catering for. A hotel is easy enough to make. At the simple end, get a bit of plumbing pipe a few inches wide and a few inches long and stuff it with twigs; or for a nicer look, make a wooden box with an open front and fill each subdivision with bits of wood, logs, or twigs. Each different type of filling will attract different insects.

Trees with rough bark or shedding bark and the occasional rotting log also provide habitat for beneficial insects and can attract insect-eating birds, as well.

After all this, do you still think aphids are a 'pest'? Yes, they are, but they're also the best way to attract some of these beneficial insects. Think of yourself just sacrificing a very small percentage of your plants to the aphids in order to give your ladybugs and hoverflies a gourmet meal.

DISTRACT BAD BUGS!

By the way, be careful about the companion plants to your crop. Most insects are quite choosy about what they eat. They have evolved to eat just a single plant family; the shape of their jaws, for instance, may have evolved to adapt to a single plant. Cabbage white butterflies aren't interested in laying eggs on your tomatoes. But they may have identified weed plants that they're happy eating or that usually grow near the plants they like.

So planting a distraction garden isn't a bad idea. Sunflowers, nasturtiums and pansies planted in a far corner might keep quite a few of the bad bugs from finding out where you've hidden your vegetable patch. And planting strong-scented herbs like sage, rosemary or mint near your vegetables can also help deter those bugs which use scents to orientate themselves from getting too close to your tomatoes and salad leaves.

BIRDS IN THE PERMACULTURE GARDEN

I love watching birds in my garden. We have a cheeky little robin who hops around and always waits when I dig up root vegetables to see if there are any worms for him; blackbirds who weave liquid melody around us in the early morning; and hosts of tiny birds who flock to the rose hips in autumn. From time to time, a woodpecker comes to look for insects in the mulch and a couple of jays do aeronautics above the orchard.

But they can be pests. They can clear a cherry tree in half an hour. I often have to share my figs; the blackbird gets a quarter and I get the rest! So a balanced permaculture garden needs to encourage your birds to do more good than harm.

They can eat pests. Thrushes love eating slugs and snails; they'll smash snails against a rock to get the soft flesh out and you can sometimes find these rocks surrounded by fragments of snail shell. A family of chickadees will eat over twenty million insects in a year. And birds can also eat weed seeds, stopping the weed from spreading.

They also leave manure. If you're smart, you'll arrange perches and bird-feeders over places where the soil needs a bit of help—it's a way of getting a bit more fertility into the soil.

Some birds, particularly chickens and related species, will scratch the soil, tilling it and reducing weeds. You may need to move them around, though, to stop them baring the soil completely and letting it dry out and erode. Some small birds will even pollinate plants for you; if you live anywhere with hummingbirds, which sip nectar through their long beaks, they will help with pollination.

So how do you attract birds? Think about what they need. Birds need water that is shallow enough for them to bathe in as well as drink. Watching birds take a bath can be one of the most amusing ways to spend five minutes; there are splashers, dunkers and divers, besides the I'm-afraid-to-get-my-toes-wet birds, which try to get water all over their wings without actually going in the water. A pond or stream may be better than a bird-bath, which could freeze in winter (though if you pour boiling water in every morning, it will defrost for a while).

Birds also need food. Depending on the bird, that might mean fruit, insects, or nuts, grubs and larvae in the soil or in deadwood, or seeds. You need to provide something for every kind of bird at all times of the year so that you create food and habitat diversity. The more diverse and complex a habitat you create, the more birds can live there.

Insectary plants don't just attract insects; they will also attract birds who eat insects, getting two functions out of one plant. Winter fruits, nuts and acorns will make your garden a particularly welcome refuge over the late fall and winter. For seed-eaters, evergreen trees, grasses, flowers and herbs can all provide food. Neat gardens where the flowers are all 'dead-headed' after blooming are far less attractive than gardens where the seed heads have been left to mature.

Particular fruit and seeds that birds love are *pyracantha* (firethorn) and *berberis* (barberries), mistletoe, holly, ivy, rosehips and hawthorn (fruits and haws late in the year), teasels, sunflowers and thistles. Not all these plants want to be in zone 1; some of the thornier plants I would probably want to put in zone 3 or 4.

Over winter, using bird feeders is a good idea. Again, try to create diversity by providing different types of food—some birds eat larger seeds, others only the smallest seeds. Distribute your feeders around the garden, rather than putting them all in one place. And keep them clean; give them a good wash every so often to avoid avian diseases being passed on.

Birds also need shelter and protection. A windy garden where flying is hard and where smaller birds can't take shelter if a sparrow-hawk or buzzard appears won't attract them nearly as much as a garden with trees for shade, protection and perching in. Even birds who will come out onto a lawn or bare earth often like to know that there's a hedge or tree nearby where they can fly to if a predator (or a gardener!) appears.

Climbers such as honeysuckle, ivy, clematis and vines will appeal to many birds. Some also like long grass that they can shelter in, while others like short grass where they can pick to find worms, so though most permaculture books won't tell you that you need a lawn, I think a small one is not a bad idea at all.

You may even be lucky enough to attract birds to nest. Some prefer to nest in trees, others in hedges, others in cracks in walls or under the eaves of buildings (e.g., swallows). Providing nest boxes can attract birds, though don't put too many too close to each other—four is enough for most suburban gardens. Depending on the size of the hole in the front, they will attract different sizes of bird.

A nesting box for titmice and other small birds is easy to make out of larger branch prunings. Drill a hole from one end of the branch and hollow it out a little more with a rasp, then drill an access hole in one side and use a small slice of the branch as a lid. You could also pay for plywood to make nesting boxes or go the whole hog and buy a nesting box. Up to you!

Hanging up old teapots (spout down, for drainage) sounds a bit crazy, but it attracts many small birds like robins. If you have cats, try to keep your nesting boxes, birdbath and any feeders cat safe; hang the feeders high and away from the house; get a birdbath on a plinth with a large overhang.

'Fair shares' can apply to birds. Let them have crab apples, fallen fruit and some of your blackberries and elderberries; if you grow globe artichokes and let one or two go to seed—the birds will thank you for it. Sunflowers are popular with many birds too.

If you have a local wildlife trust, they may have information available for you and even know sources of free plants. One local association helped me create an owl nesting box for the side of the house; I haven't seen the owl yet, but I have heard her and she leaves plenty of pellets underneath.

HOW TO MAKE A NESTING BOX FROM BRANCH PRUNINGS

The birds and the bees may be the most important wildlife in your garden, but there are some other beneficial animals too. Frogs and toads eat slugs and insects (including mosquitoes), so building a frog pond is a great idea. It will also attract birdlife and dragonflies, both good pest controllers. Amphibians like dappled sunlight and lots of nooks and crannies, like the gaps between rocks or rotting wood. Keep the pond shallow; just ten centimeters is enough for frogs and tadpoles (have another separate pond if you want goldfish or koi).

BIRD FEEDING FROM A FRUIT TREE

BIRD ENJOYING A BATH IN A GARDEN POND

Beneficial plants and trees for birds						
Plant/tree	Seeds	Food	Autumn/winter food	Insects	Shelter	Nesting site
Alder	•					•
American beautyberry	•	•	•			
American holly		•	•		•	•
Apple tree		•	•		•	•
Ash	•					•
Bamboo					•	•
Barberry					•	•
Birch	•			•		
Berberis		•	•		•	•
Black eyed susan	•					
Blackberry			•			
Blueberry			•			
Buckthorn					•	•
Butternut		•		•	•	•
Cherries		•				
Chestnut		•		•	•	•
Coral honeysuckle		•				
Cotoneaster		•	•			
Crabapple		•	•		•	•
Currants		•				
Dog rose		•				
Dogwood		•	•			•
Douglas fir tree					•	•
Drummond red maple	•				•	
Eastern red cedar tree			•	•	•	•
Elderberry		•			•	•
Elm tree				•		•
Euonymus	•					
Fir tree					•	
Goji berry		•	•		•	•

Beneficial plants and trees for birds						
Plant/tree	Seeds	Food	Autumn/winter food	Insects	Shelter	Nesting site
Grasses	•				•	
Greenbrier					•	
Guelder rose		•				
Hardy kiwi		•				
Hawthorn tree		•	•		•	•
Hazelnut		•		•	•	•
Hickory		•		•	•	•
Holly		•	•		•	•
Honeysuckle		•				
Ivy		•	•		•	
Junipers		•	•		•	•
Maple tree	•			•		•
Mulberry		•			•	
Oak tree		•		•	•	•
Persimmon		•	•		•	•
Pine tree					•	
Pyracantha		•			•	•
Raspberries		•				
Red mulberry		•				
River birch	•	•			•	
Rose		•	•		•	•
Rowan		•				
Salal		•			•	•
Serviceberry		•			•	•
Southern magnolia	•				•	
Southern wax myrtle		•	•		•	
Spicebush		•				
Spindle		•	•			

Beneficial plants and trees for birds						
Plant/tree	Seeds	Food	Autumn/ winter food	Insects	Shelter	Nesting site
Spruce tree	•				•	•
Strawberries		•				
Sunflower	•					
Sycamore				•		
Teasel	•					
Virginia creeper			•	•		
Walnut			•		•	•
Whitebeam			•			
Willow				•		
Wolfberry		•			•	•

SMALL ANIMALS IN THE PERMACULTURE GARDEN

What I'm going to talk about here is not small animals like mice and raccoons that may decide to utilize your garden, but livestock that you introduce deliberately—chickens, geese, turkeys, ducks, rabbits, or guinea pigs. They can all be quite useful in several ways. All of them drop manure, which is good for your garden; chickens will scratch the ground and eat up pests; all of them are also happy to eat scraps and leftovers. Some, particularly geese, will also act as guards, letting you know if anyone approaches the house.

If you only have a small garden, you probably don't need to read the rest of this chapter; besides, if you live in a suburb, your neighbors may not want a rooster waking them at dawn on Saturday morning! But for larger permaculture gardens, small livestock can help keep the garden under control and reward you with manure, eggs, meat, or companionship (I think it's usually meat *or* companionship, not both).

You decide where they go; otherwise, it's a recipe for disaster. For instance, with chickens, if you have a zone 3/4 area of food forest with mainly taller plants, they will keep the forest floor weed and pest free. But you don't want to let them in your vegetable bed! As for goats, they need to be kept out in zones 4 and 5; they can (and will) eat just about anything; they can also climb and jump, so you need to be rigorous about fencing.

A 'chicken tractor' is a great idea for keeping your chickens under control. It's basically a big chicken pen on wheels, so you can decide to put your chickens in one place till they've got rid of all the undergrowth and slugs, then wheel it onto a new patch that needs the chickens' attention. The chickens will run along with you! On the patch that the chickens have cleared, you can then plant a cover crop and chop it down when it's ready to provide a fertilizing mulch.

Chickens will also need protection from certain animals. For example, in the UK, foxes will try any way they can to get into a chicken pen, so you may need to use electric fences to stop this.

Keeping chickens (or rabbits) too long in one place can lead to soil compaction, over-scratching and destruction of vegetation. The chicken tractor is an excellent way of getting the benefits without the downside. You can use a similar approach with a 'rabbit tractor' too, but because rabbits can dig, you need to put chicken wire on the bottom of the tractor as well as the sides.

You can also let livestock into the vegetable bed once you've harvested. 'Fair shares' again—if you've eaten all the tomatoes and corn and harvested the squashes for keeping over winter, let the animals in. They'll clean things up ready for you to mulch or plant the next crop.

Remember that livestock takes more looking after than plants. Chickens may feed themselves, but they do need a house that's got roosting perches and is safe from predators. Our neighbors have chickens, who roam most of the garden during the day (a small part is gated off) and make their way back to the gate of their fenced chicken compound in time for their owner to lock them in before night comes (the house also provides a place where the hens can lay their eggs—they will lay anywhere, but the advantage of the chicken house is that you know exactly where to look for them).

Ducks can be stupid and just as noisy as chickens, but they have the advantage of eating slugs and snails and they are tougher and hardier than chickens. They're also fun to watch. Unlike chickens, they're fine to leave with established plants, but don't leave them with seedlings.

By the way, ducks are very social, so you need at least two. Indian runners have great characters and are real garden ducks, but they're not the most prolific layers. However, their eggs are very tasty.

Geese can be noisy and aggressive, so they're not for a suburban garden. Get young goslings and tame them if you want to enjoy their company. Geese aren't interested in your fruit and veg; they nibble grass, so they're a permaculture lawnmower. But they produce a lot of manure. Too small a garden and too many geese can get very nasty indeed.

If you're not happy with domesticating animals (many vegans aren't), or if you simply don't have the time for livestock, then it is fine to run a permaculture garden without them. Chicken manure is a really great fertilizer, but you can get along without it, or maybe a local farmer will provide you with some for free. You can also buy the pelleted variety. Wildlife will give you some of what you need; you may just have to work a little harder mulching. Instead of rotating grazing or chickens, you'll rotate green manures and other crops.

A vegetarian friend of mine has a different outlook. She rescues old battery hens that would otherwise be killed as non-productive and lets them run around her orchard. They eat her kitchen peelings, they control pests and they provide heaps of excellent manure. If they lay eggs as well—some do, some don't—that's a bonus. And they have a good, happy life.

EXAMPLE OF CHICKEN TRACTOR

EXAMPLE OF A CHICKEN TRACTOR

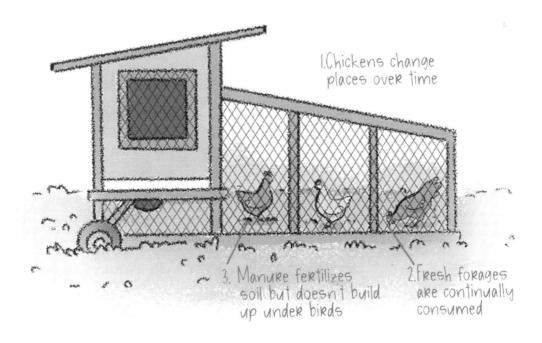

1. Chickens change places over time

2. Fresh forages are continually consumed

3. Manure fertilizes soil but doesn't build up under birds

CHICKEN TRACTOR DESIGN
TAKEN FROM WWW.FIX.COM

BEEKEEPING

If you have ever had an allergic reaction to a bee sting, don't even think about beekeeping. Being stung can bring about anaphylactic shock. However, for most of us, beekeeping is a good way of fulfilling multiple functions—pollinating our flowers and fruit and providing ourselves with an organic sweetener.

You'll need a suitable location for your hives, with sun (afternoon shade if you're in a very hot climate), freshwater (such as a pool with shallow sides or stones for the bees to stand on), shelter from high winds and privacy. Keep your hives well away from where people pass or put the hive entrance up against a wall or hedge, so the bees have to start their flight by going up high, far above the human zone. If possible, hives should also face south—and keep them above the ground.

You'll want to get your bees in spring. Professional beekeepers will usually include education as part of the package when you buy your first bees, teaching you how to use the smoker and protective clothing and how to check your bees' health. They'll be able to tell you the signs of the most common diseases in your area and they are your first port of call if your bees become diseased or don't thrive.

Although bees make honey when they get started in a new hive, making a 'nectar' from equal parts of water and sugar will help give them the energy they need to adapt. Pretty soon, they'll stop drinking so much of it and start taking nectar from flowers instead; at this point, you can stop feeding them.

A big part of beekeeping is leaving the bees alone as much as possible. Check a hive occasionally to ensure the queen is healthy—the sign is plenty of larvae, all in different stages of development. Smoke the hive to make the bees drowsy before you go in.

In the first year, your bees probably won't give you much honey. In the second year, you'll be able to start extracting it—if you don't have a professional extractor, then you just cut and crush the honeycomb, strain it, let it settle in a bucket with a tap at the bottom, and take the honey off after a couple of days. You can then render the wax, which is excellent for furniture polish, candles and beauty products like hand salve.

Beekeepers in most countries have guilds or societies which can provide you with a lot of help. You should definitely join—they'll also keep you on top of any regulatory requirements.

Once you become experienced, you'll be able to get new hives started for free by dealing with wild swarms. People often find a swarm of bees has invaded their mailbox or the eaves of their house—go out with your smoker and you can just usher them into a bee box and take them home.

EXERCISES

Which birds are missing?

Find or create a list of birds that are common in your area. Now look and see which birds are missing.

Why are they not coming to your garden? Waders, ducks and other waterfowl won't visit if you don't have a pool and shoreline waders may not come so far inland. But there may be species that you'd expect to see, but they're not coming. Look up their lifestyle and see if there is something missing—perhaps they are particularly attracted by certain berries or have very specific nesting needs.

You may now have some good information on exactly which changes to your garden could increase the diversity of birds visiting it the most.

Insect identification

You're going to need either a little book of insects or a good app on your smartphone for this. Go out into your garden any time that's not the middle of winter and watch for insects. Can you identify them? How many species are there? Also, note which plant you found them on or near.

What functions does each insect fulfill? Do they make nests, like bees? Do they have predators, or do they prey on other insect species?

Were any particular plants extra-attractive to the bugs or that showed signs of having been eaten by bugs or caterpillars?

9

BUILDING RELATIONSHIPS IN THE GARDEN

Did you find the last couple of chapters easier going? We've been talking about plants, talking about animals and insects, and not principles and theories and systems and processes.

But remember, the permaculture garden is all about relationships. If you imagine a poker hand—just seven different cards from the pack—it may not say anything at all to you. But a poker player will see it a certain way and will see the different relationships that the cards make with each other—pairs or threes of the same number card, runs (four-five-six-seven), or cards of the same suit. When that player draws another card, they'll be looking at all of those relationships. They want to see which relationships are strengthened, and which are weakened, by the new card; does the king of hearts work better with the pair of kings they've already got, or with the ten-jack-queen of hearts? So, in the same way, every time you put a new element in a permaculture garden, you're thinking about its relationship with all the others. Fortunately, unlike a poker player, you can take a new card, so to speak, without having to throw one out!

Permaculture isn't just gardening; it's a system of ecological design, an art of designing relationships.

We design the relationships between plants and other plants, between plants and the soil, plants and animals, and the whole garden and ourselves.

I should point out that what we know about these relationships is always evolving. Sometimes, science shows that traditional companion plantings don't really work; sometimes, it shows us new possibilities for creating relationships, for instance, between fungi and plants. When I say 'science,' I don't just mean university research scientists working in the lab; many permaculture farms and institutes keep detailed records of their plantings to assess how well each variant works. You should do the same. Keeping good records can be a hassle for the first couple of years, but after that, you will have a real resource to look back on and learn what works best in your particular ecosystem.

INTERPLANTING

Interplanting, also sometimes called *intercropping*, is occasionally used by conventional gardeners too. Growing basil and tomatoes in adjacent rows makes sense if you're going to want to pick some basil every time you make a tomato salad; and the two plants don't compete, because the basil is low and the tomatoes are tall, and the tomatoes like the sun while the basil likes a bit of shade. In fact, permaculture farmer Toby Hemenway called tomato, basil, and pepper the "spaghetti sauce guild." You can see exactly what he was thinking there!

But in permaculture, we take it a bit further. Interplanting is a good idea for several reasons; we're making better use of space, we can make better use of time, and sometimes, the plants will have a beneficial relationship or a functional relationship that makes our garden more productive.

For instance, let's make use of time. We can plant a fast-growing crop between slower-growing plants, like lettuce between broccoli. Lettuce likes shade, and the broccoli will shade it; by the time the broccoli has grown really tall, you'll have eaten the lettuce. Or we can make use of space, such as by planting a short crop between tall plants; for example, lettuce or herbs between beans.

Another intercropping that works well is sowing mixed salad leaves like lettuce, arugula, or mustard around tomato seedlings. The salad will stop weeds growing near the tomato, and as you gradually pick the leaves, the tomato grows. You've probably eaten all the baby salad by the time the tomato wants to go it alone.

You may remember the example of lettuce mixed up with some radish seed. The radish comes up really quickly, so you don't lose sight of where you sowed the lettuce, and as you eat the radishes, you're getting the effect of thinning out the lettuce seedlings without wasting any of them. We could even go a step further and add sweet peppers as another component—by the time the pepper plants have grown to full size, you've already eaten the radish and the lettuce.

RADISH, SWEET PEPPER, LETTUCE

Radishes and carrots work well; the radishes will be ready in three to five weeks, while carrots take three months to ripen.

RADISH AND CARROTS

Cabbage and cauliflower don't bulk out for a couple of months, so that gives you time to intercrop spinach or beets between them.

Bush beans, on the other hand, are fairly quick to grow, so put them between tomatoes or peppers. This particular pairing has another benefit, which is that they also fix nitrogen, which tomatoes need.

CABBAGE, BEETS, BUSH BEANS AND TOMATOES

With all these intercroppings, you never leave any soil bare; there's always a plant growing, and its leaves stopping the wind and rain from getting at the soil.

Remember that you can stack crops vertically, even in a raised bed, using tall crops to shade smaller crops. Tomato and corn love the sun, and tomato will do a good job of shading salads or beet. You can also grow squashes, cucumbers, zucchini and melons on trellises, taking them up vertically to save space and shade other vegetables at the same time. Shade-lovers include arugula, endive, lettuce, mustard, pakchoi and tatsoi, chard, radish, mizuna, and beets.

Intercropping

Fast growers

Arugula, bush beans, carrots, radish, mizuna, green onions, tatsoi, broccoli raab, beets

Cilantro, dill, annual herbs

Slow growers

Broccoli, cabbage, brussels sprouts, cauliflower, tomatoes, corn, kale

Grow tall

Tomato, corn, squashes, cucumbers, zucchini, melons

Like shade

Salads, beets, arugula, endive, lettuce, mustard, pakchoi, tatsoi, chard, radish, mizuna

The famous 'three sisters' planting is another use of vertical stacking; however, it doesn't work so well in the UK as in the US, as the corn doesn't grow quite high enough. I still use it, but I use a couple of poles to help the beans, too. Maybe just two sisters, beans and squash, would work better.

In this guild, the corn grows tall and acts as a support for the pole beans, which fix nitrogen in the soil for the hungry squash. The squash sprawls out over the soil, acting as a mulch to keep the moisture in for the two other plants' roots. It's an elegant system, as well designed as clockwork or good software.

You can intercrop in rows or relays, but you can also plant a mixed intercropping, for instance, by using concentric circles in a keyhole bed. Or you could zigzag (which is actually an effective way to use space because you can get more vegetables in a bed that way), or you can plant down the middle and then along the sides of a bed. Don't feel you have to use straight lines—remember when we talked about natural patterns, straight lines didn't feature at all.

To intercrop successfully, you need to think about root systems. Some plants, like most salad leaves and potatoes, have shallow roots; carrots, beets, tomatoes, and parsnips have deep roots.

You'll also want to think about leaf shapes. For instance, onions, carrots, and lettuce work together because they have different leaf forms; onions are long and thin, carrots feathery, lettuce big and floppy. They have different light requirements, lettuce liking a little bit more shade, but neither onions nor carrots throw too much shade, owing to the shape of their leaves. And their roots don't compete, either; onions have shallow roots, lettuce has a network of small roots that go down a little further, while carrots have taproots reaching much further down.

A few more intercroppings that work are:

- Peas, lettuce, cabbage.
- Radish, lettuce, pepper.
- Broccoli and peas.
- Cabbage-pumpkin-marigold.
- Basil-asparagus-tomato.

Intercropping just to save space works, but as the example of the three sisters shows, you can intercrop plants in order to achieve a dynamic relationship between them. The plants give each other mutual benefits: a plant can deter another plant's pests or give the other plant nutrients. Now we've moved on from intercropping to companion planting—from people just sharing a house as a place to live to people running a communal life and helping each other with their different skills and abilities.

You may also have noticed that a lot of this intercropping runs slightly counter to the permaculture ethic because it's all about annual plants, whereas permaculture prefers to use perennials and self-seeders. But that's a compromise we make, to get the amount of food we need.

COMPANION PLANTING

Companion planting is an aspect of permaculture that's long been part of gardeners' mythology, accepted even by quite conventional gardeners. But Robert Kourik showed in his book on *Edible Landscape* that some of the long-standing companion plantings may just be myths, with no scientific reasons behind them.

For instance, the idea that marigolds deter pests may be flawed. It is known that marigold can deter nematodes and whiteflies, but it was often thought that they could deter an array of pests, with Mexican or French marigold proving the most useful. There is no conclusive scientific evidence, but it has often been thought that marigold can also deter pests such as hornworms, cabbage worms, thrips, squash bugs and others. But many gardeners still use marigold with other plants, as it is thought that their scent alone may deter pests. They are also a trap for slugs, keeping these pests away from other plants!

However, we reproduce these companion planting ideas for two reasons. First, because permaculture tries to copy things that work in nature, and many of these plantings are based on what we find in forest and woodland or meadowland that hasn't been gardened or 'managed.' Second, because you should be keeping good records and so you should evaluate each of these plantings and how it works for you, on your particular site.

Companion planting will help the biodiversity of your garden, too. Remember that biodiversity isn't just about having different species but about different relationships between the species. The more companion plants you have, the more relationships between them, and the more resilient your garden.

Allelopathy can be beneficial or harmful, but the main instance you should be aware of is the black walnut, which prevents other plants from growing around its roots. Its leaves contain a growth-suppressing chemical called juglone. When the leaves fall, they create a toxic mulch under the tree, which prevents other plants from competing for nutrients with the walnut's roots. Tomatoes, peppers, aubergines (eggplants), and potatoes, in particular, will not grow in the shade of a walnut. By extension, you should not use walnut leaves in your leaf mold or mulches.

Some other allelopathic plants include English laurel (*Prunus laurocerasus*), sumac, rhododendron, forsythia, Kentucky bluegrass, and elder.

The following are companion planting combinations you might try:

- Sage near carrots repels carrot fly.
- Swiss chard and sweet alyssum looks good and is good. Chard is a great cut-and-come-again crop and comes in red, orange, and 'rainbow' varieties, contrasting with the white alyssum. Chard's long taproot doesn't interfere with the alyssum's fibrous, shallow roots. And the chard has lots of vitamin C for your diet. The alyssum also provides a living mulch for the chard and attracts beneficial insects. So this is actually a relationship in which there's a lot more going on than just saving space.
- Asparagus and tomatoes are mutually beneficial; tomatoes protect against asparagus beetles and asparagus protects against some tomato pests.
- Nasturtium and rosemary will deter bean beetles.
- Borage repels tomato worms. It also attracts bees, so it's a good pollinator-attractor.
- Rosemary and thyme—indeed any heavily scented herbs—planted with any brassica can help to deter cabbage fly.
- Dill can improve the growth of your brassicas. Some say it also enhances the flavor.
- Clover can help your cabbages. It increases the number of predator ground beetles that will eat cabbage worms and aphids. Mint and sage will deter cabbage moths.
- Chives are useful; planted around fruit trees, they deter insects from climbing the trunk; they deter aphids and can help the growth of carrots and tomatoes.
- Garlic is another helpful all-purpose plant (except with beans). It accumulates sulfur and discourages many pests, including aphids and flea beetle.
- Horseradish is perhaps not a vegetable you're going to use a lot, but it can keep pests away from potatoes and plum trees. And it's easy to grow—in fact, the problem is more likely to be keeping it under control. Being British, of course, I love horseradish with my roast beef and I've never had a shortage of horseradish!
- Potatoes are vulnerable. Don't plant cucurbits or tomatoes too close or raspberries. All these species are vulnerable to verticillium wilt, and if one gets the disease, the others will get it too. On the other hand, horseradish can help disease and pest resistance.

- Sweetcorn needs huge amounts of nitrogen, so plant it with nitrogen fixers—as in the 'three sisters'—such as beans and peas.
- Radishes can deter cucumber beetles.
- Nasturtiums under fruit trees help discourage pests; they also deter aphids and other cucumber-eating pests.
- Chili peppers have strong roots and can help cure root rot in other plants.
- Strawberries will benefit many plants; borage tends to deter many of the insect pests that afflict strawberry plants and are a powerful pollinator attractor and can make pollination more robust to help new strawberry plant variations grow. Caraway attracts parasitic wasps and parasitic flies which benefits the strawberry plants. Lupin is a nitrogen fixer and also attracts bees.

The following are companion planting combinations you might want to avoid:

- Corn and tomatoes are a mix with a curse; the same pests eat both plants. Plant them together and you're sticking up a huge "Free Food Here" sign for the tomato and corn earworms.
- Potatoes, tomatoes, and peppers are all in the same family and are all susceptible to blight and other shared diseases, so plant them well apart to reduce the risk of cross-contamination.
- Strawberries and cabbages or Brussels sprouts make a bad mix. Broccoli and cauliflower don't go well with
- Keep fennel away from your annual crops—it may be better off in a wilder part of the garden, not in the vegetable bed because it is allelopathic to most garden plants and can kill them.
- Pole beans inhibit the growth of beetroot, so they do not make good companions.
strawberries either. Strawberries will impair the growth of these plants.
- Cucumbers and potatoes don't grow well together, as the cucurbits can encourage blight.

Companion planting

Plants	Good companions	Bad companions
Apples	Chives, horsetail (equisetum), foxgloves, wallflowers, nasturtium, garlic, onions	Grass, potatoes
Anise	Coriander	Basil, rue
Apricots	Basil, tansy, southernwood	Tomatoes, sage
Asparagus	Tomatoes, parsley, basil	
Basil	Tomatoes, asparagus, parsley, apricots, sweet pepper, capsicum, chilli, fuchsias, grapevines, peaches	Rue, anise, rosemary, sage
Beans	Carrots, cucumber, cabbage, lettuce, peas, parsley, cauliflower, spinach, summer savory, sweetcorn, squash, celery, eggplant, marigold, parsnips, potatoes, radish, strawberries, turnips	Onions, garlic, fennel, gladioli, sunflowers, kohlrabi, beetroot, chives
Beans, Dwarf	Brassicas, carrots, cucumber, dill, lettuce, potatoes, radish, spinach, sweetcorn, strawberries, summer savory, beetroot, brussels sprouts, broccoli, capsicum	Onions, garlic, fennel, gladioli, sunflowers, kohlrabi
Beans, haricot	Carrots, cucumber, parsley, lettuce, peas, brassicas (though not for bush beans)	
Beetroot	Onions, silverbeet, kohlrabi, lettuce, cabbage, dwarf beans, spinach, cauliflower, brussels sprouts, broccoli, radish	Tall beans, mustard
Borage	Strawberries, fruit trees, tomatoes	
Brussels Sprouts	Beans (dwarf), beetroot, celery, cucumber, onions, rhubarb, chamomile, dill, oregano, sage, marigold, nasturtium	Strawberries
Broccoli	Beans (dwarf), beetroot, celery, cucumber, onions, rhubarb, chamomile, dill, oregano, sage, marigold, nasturtium, lavender, lettuce, silverbeet	Strawberries
Cabbages	Beans, beetroot, celery, mint, thyme, sage, onions, rosemary, dill, potatoes, chamomile, oregano, hyssop, southernwood, nasturtiums, tansy, coriander, wormwood, peppermint, spinach, clover, lavender, lettuce, silverbeet	Rue, strawberries, tomatoes, garlic, mustard
Calendula	Chard, radish, carrots, tomatoes, thyme, parsley, valerian	

Companion planting		
Plants	Good companions	Bad companions
Capsicum (sweet Peppers)	Amaranth, basil, beans (dwarf), carrots, lovage, marjoram, okra, parsley, geraniums	Fennel, kohlrabi, tomatoes
Carnations		Hyacinths
Carrots	Peas, radish, lettuce, chives, sage, onions, leeks, rosemary, wormwood, parsley, beans, calendula, capsicum, fruit trees, tomatoes, turnips	Dill, parsnips, caraway, fennel
Cauliflower	Celery, spinach, beets, beans, tansy, nasturtiums, silverbeet	Strawberries
Celeriac	Scarlet runner beans	
Celery	Tomatoes, dill, beans, leeks, cabbage, cauliflower, brussels sprouts, broccoli, chard, cucumber, spinach	Lettuce, parsnips
Chamomile	Mint, cabbage, onions, brussels sprouts, broccoli	
Chard	Roots crops, lettuce, radish, celery, mint, alyssum, calendula	
Chervil	Dill, coriander, radish	
Chilli (hot Peppers)	Basil, oregano, parsley, rosemary	Potatoes, tomatoes, eggplant
Chives	Parsley, apples, carrots, tomatoes, fruit trees, rosemary, silverbeet, turnips	Peas, beans
Citrus	Guava	
Collards	Tomatoes	
Coriander	Dill, chervil, anise, cabbages, carrots, radish, chard	Fennel
Cucumbers	Potatoes (early crop only), beans, celery, peas, lettuce, sweetcorn, savoy cabbages, sunflowers, nasturtium, okra, radish, brussels sprouts, broccoli, rhubarb	Potatoes (with any but early crop), aromatic herbs, rue, sage, savory, silverbeet
Dill (beneficial companion plant to tomatoes when it is young, mature dill plants can suppress growth of tomatoes though)	Tomatoes (young dill), cabbage, fennel, coriander, brassicas, beans (dwarf), brussels sprouts, broccoli, celery, chervil	Carrots, caraway, fennel, tomatoes (mature dill)

Companion planting		
Plants	Good companions	Bad companions
Eggplant	Beans, okra, spinach	Potatoes, tomatoes, peppers, chilli pepper
Fennel		Beans, tomatoes, kohlrabi, coriander, wormwood, most annuals, capsicum
Foxgloves	Apples, potatoes, tomatoes	
Fruit Trees	Chives, garlic, carrots, bulbs, borage, strawberries, nasturtium, comfrey, plantain, columbine, daylilies, horseradish	
Fuchsias	Basil, gooseberries, tomatoes	
Garlic	Roses, apples, peaches, tomatoes, fruit trees, parsnips, silverbeet	Peas, beans, cabbage, strawberries
Geraniums	Grapevines	
Gladioli		Strawberries, beans, peas
Grapevines	Geraniums, mulberries, hyssop, basil, tansy, peaches	
Guava	Citrus	
Horseradish	Fruit trees, potatoes	
Hyacinth		Carnations
Hyssop	Grapevines, cabbage	Radish
Jerusalem Artichoke	Sweetcorn	
Kohlrabi	Beetroot, onions, nasturtium	Tomatoes, beans, fennel, capsicum
Lavender	Broccoli and cabbage family, rosemary, silverbeet	
Leeks	Carrots, celery, onions, silverbeet	
Lettuce	Carrots, onions, strawberries, beetroot, cabbage, cucumber, radish, marigold, broccoli, tomatoes, sweet potato, beans, chard	Parsley, celery
Marigold	Lettuce, potatoes, tomatoes, roses, beans, brussels sprouts, broccoli, rosemary, sweetcorn	
Melons	Sweetcorn, okra, radish	Silverbeet
Mint	Cabbage, chamomile, chard	Parsley, rosemary
Nasturtiums	Apples, cabbage, cauliflower, broccoli, brussels sprouts, kohlrabi, turnips, radish, cucumber, zucchini, rosemary, fruit trees, potatoes, squash, tomatoes	

Companion planting		
Plants	Good companions	Bad companions
Nettle	Increases oil content of most herbs	
Okra	Melons, cucumber, sweet peppers, eggplant, capsicum	
Onions	Carrots, beetroot, silverbeet, lettuce, chamomile, tomatoes, kohlrabi, summer savory, roses, apples, brussels sprouts, broccoli, cabbage, leeks, spinach	Peas, beans, sage
Oregano	Cabbage, brussels sprouts, broccoli, chilli, rosemary	
Parsley	Tomatoes, asparagus, roses, chives, carrots, basil, beans, calendula, capsicum, chilli	Potatoes, mint, celery, lettuce
Parsnips	Peas, potatoes, sweet pepper, beans, radish, garlic, silverbeet	Carrots, celery, caraway
Peaches	Tansy, garlic, basil, southernwood, grapevines	
Pears		Grass
Peas	Potatoes, radish, carrots, turnips, cucumber, sweetcorn, beans, aromatic herbs, parsnips, spinach	Onions, garlic, shallots, chives, gladioli
Peppers (sweet)	Basil, okra, parsnips	Potatoes, tomatoes, eggplant, chilli pepper
Potatoes	Peas, beans, cabbage, sweetcorn, broad beans, green beans, nasturtium, marigold, foxgloves, horseradish, cucumber (early potato crop), parsnips	Apples, cherries, cucumber (with any but early crops), pumpkins, sunflowers, tomatoes, raspberries, rosemary, squash, peppers, silverbeet, turnips, chilli pepper, eggplant
Pumpkins	Sweetcorn, datura, pole beans	Potatoes, rosemary
Radishes	Lettuce, peas, chervil, nasturtium, cucumber, beets, spinach, carrots, squash, melons, tomatoes, beans, calendula, chard, parsnips, sweet potato	Hyssop
Raspberries	Tansy, rue	Blackberries, potatoes
Rhubarb	Columbines, brussels sprouts, broccoli	
Rosemary	Chives, brassicas, strawberries, oregano, thyme, sage, alyssum, carrots, marigold, lavender, nasturtium, cabbage, chilli	Mint, basil, pumpkins, tomatoes, cucumbers, potatoes

Companion planting		
Plants	Good companions	Bad companions
Roses	Garlic, parsley, onions, mignonette, marigolds, rue, tansy, tomatoes	
Rue	Roses, raspberries, fig trees	Sage, basil, cucumbers, cabbage, anise
Sage	Carrots, cabbage, strawberries, rosemary, brussels sprouts, broccoli	Basil, rue, wormwood, cucumber, onions, apricots
Savory	Beans, onions	Cucumber
Silverbeet	Beetroot, parsnips, tomatoes, lavender, brassicas (broccoli, cabbage, cauliflower, etc.), onion family (chives, garlic, leeks, onions, etc.)	Potatoes, sweetcorn, cucurbit family (cucumbers, gourds, melons, squash, etc.), most herbs
Spinach	Broad beans, cabbage, beets, cauliflower, celery, eggplant, onion, peas, strawberries, santolina (cotton lavender), radish, turnips	
Squash	Sunflowers, sweetcorn, clover, nasturtium, radish, beans	Potatoes, silverbeet
Strawberries	Borage, lettuce, spinach, sage, pyrethrum, beans, caraway, lupin, fruit trees, rosemary	Cabbage, cauliflowers, brussels sprouts, gladioli, tomatoes, broccoli, garlic
Sunflowers	Squash, cucumber, sweetcorn	Potatoes, beans
Sweet Potato	White hellebore, lettuce, radish	
Sweetcorn	Potato, peas, beans, cucumber, pumpkin, squash, melons, marigolds, sunflowers, sunchokes, tansy, broad beans, Jerusalem artichoke	Tomatoes, silverbeet
Tansy	Cabbage, roses, raspberries, grapes, peaches, apricot, cauliflower, sweetcorn	
Thyme	Cabbage family, rosemary, calendula	
Tomatoes (dill is beneficial companion plant to tomatoes when it is young, mature dill plants can suppress growth of tomatoes though)	Asparagus, celery, parsley, basil, carrots, chives, marigold, foxgloves, garlic, onions, nasturtium, roses, bee balm, borage, lettuce, mustard, calendula, collards, dill (young dill), fuchsias, radish, silverbeet	Rosemary, potatoes, kohlrabi, fennel, apricots, strawberries, dill (mature dill), cabbage, sweetcorn, peppers, capsicum, turnips, chilli pepper, eggplant
Turnips	Nasturtium, peas, beans, carrots, chicory, chives, spinach	Potatoes, tomatoes

Companion planting		
Plants	Good companions	Bad companions
Valerian	Calendula, echinacea	
Wallflowers	Apples	
Wormwood	Cabbage, carrots	All other plants
Zucchini	Nasturtium	

POLYCULTURAL GARDENING

Moving on from intercropping, which is mainly about annuals, and companion planting, which mixes in some of the perennials, we can go a stage further and look at polycultural gardening. This combines interplanting and companion planting in a polyculture with multiple crops layered in the same space, in an imitation of natural ecosystems. Polycultural gardens look more like a natural plant community than like an ordered vegetable bed.

They also succeed in growing a wide range of different foods by creating different niches. Most vegetable beds or monocultures have all their plants in the same conditions of soil and sun or shade. Instead, you're aiming to create different niches with varying soil, sunny and shady spots, sun traps and cooler areas, wet areas and freer-draining areas. The more niches you have, the greater the number of species your garden will support.

Diversity creates better resistance to diseases. In China, it was found that planting different varieties of rice together increased yields 89%, mainly because it almost completely prevented disease. Farmers no longer needed pesticides. Monocultures, on the other hand, are at the mercy of pathogens; if one plant is affected, the disease spreads rapidly through the whole crop.

So let's build up some polycultures. We could start with Jerusalem artichokes/sunchokes; they're tall, fast-growing plants, but it's the tubers that are edible. And then there are hog peanuts (*Amphicarpaea bracteata*) which have edible 'roots,' or rather, underground seeds. Hog peanuts are fast growers too, but they sprawl, so the artichokes will grow right up through them. And though their seeds are pretty small, since they mature at the same time as the chokes, you get a double harvest when you're digging up the artichoke tubers.

Now the hog peanut is a good nitrogen fixer, and it's a great ground cover. That's two functions in one plant, and we can put them together with berry bushes such as jostaberry (a thornless gooseberry/blackcurrant hybrid). The jostaberry gets a 'living mulch' that stops weeds growing up around it and also benefits from the nitrogen.

Or we could look at what we can do with comfrey and mint. They make a great, tall ground cover for underneath trees. Why use both? The comfrey is a dynamic accumulator; it's a good ground cover and makes a good mulch. The mint is a tasty herb and attracts beneficial insects, but it also distracts pest insects away from your edible plants.

Several polycultures can be built up with ramps—wild leeks (*Allium tricoccum*) and camas (*Camassia quamash*). Though they're not vegetables you can find in the supermarket, ramps are tasty additions to your diet and camas have edible bulbs that taste like a cross between sweet chestnut and cooked pears. They are happy to grow in the shade, so they are a reliable ground cover under trees.

Suppose we take a beach plum (*Prunus maritima*); while it's starting off, we could use dwarf coreopsis as ground cover and to attract beneficial insects, and once it's established enough to have created a good shade, add the ramps and camas. You then have a really nice little polyculture which should keep regulating itself. Maybe eventually, the coreopsis will give up, but that's okay; it's done its job. That's all part of the dynamic, self-regulating nature of a good polyculture. Camas (camassia) is just hardy enough for most of the UK. The flowers are really gorgeous and the bulbs are edible roasted, baked, steamed, or stewed.

Or you could add to the beach plum/ramps/camas mix a few more alliums, such as Egyptian leeks (*Allium ampeloprasum*) and elephant garlic, which will do well in the early years in the middle, and then later on the edges of the patch, as they need more sun, while the ramps will take over in the middle. Plants will often find their right place, and, unlike conventional gardeners who want them in neat rows, as permaculturists, we let them do it; that's where they will do best.

A polyculture will always work best if you think about the functions of the plants in it. Which plant is a good nitrogen fixer? Do you have a good ground cover? A plant with deep roots that can help break the ground up? A nutrient accumulator? Do you have plants that will attract pollinating insects, birds (which fertilize the ground with their manure), and other beneficial wildlife or deter pests?

And then, of course, you can also think about the other benefits of the polyculture. Does it provide you with food you love to eat? Does it look beautiful? Does it attract birds that you like to see in your garden? That's more subjective, but permaculture is trying to get the natural system you've set up to provide all your objectives, whether those are just productive, aesthetic, or even making your garden a wildlife haven.

Exactly which plants work together and how is still something that we're learning about. You may see very different opinions in different books or from different permaculture practitioners. So keep good records to see what works for you; choose plants to go together that don't compete for height or with the depth of their roots, and group as many as possible together for greater biodiversity.

You'll also want to pick plants for your climate. Even among, say, apple trees, grapes, or figs, you'll find some that are robust enough to grow in colder climates and others that really need heat. If you are *just* in the zone for one, but very safely in the zone for the other, pick the one that will do best; for instance, I grow Brown Turkey and White Marseilles figs, but the Brown Turkey has a great yield nearly every year, while the White Marseilles has some very poor years, because they need to be kept warm. And when I planted it, I didn't take enough care to put it in the right microclimate. If I wanted to plant another one, I'd find a sheltered place against a wall where it would get plenty of sun.

Of course, if you managed to create a microclimate, you might expand your choices. Having a south-facing, white-painted wall will often let you grow plants that need more sun and warmth than your climate zone will usually give you. Creating a backyard wetland will let you grow plants that need more water, even if you're in a fairly dry climate.

Think about whether plants are fast or slow growers. In your first seasons, you want fast growers for ground cover ('*pioneer*' plants) to stop weeds from getting established and to help open up the soil. But you might want to plant your slower-growing trees early, too, to let them get established.

Also, think about the seasonality of your plants. I have a friend whose garden is beautiful in the spring, full of daffodils, tulips and other early flowers. But there's nothing left by the time fall comes around. You need to have as long a harvesting season as possible, from your early salads and peas to your latest apples and berries. In an orchard or food forest, you will want to have some early and some late-blooming varieties of fruit trees and berry bushes. This extends your food season; it also makes sure you do get some fruit if an early frost stops your earliest blooming trees from fruiting.

In a way, putting together your polycultures is like picking a good management team. You need a finance person, you need someone with vision, you need someone who's good at keeping everyone else on track, you need someone with detailed product knowledge… and hopefully, if you put them together the right way, they'll get on with the job and support each other and you won't need to do loads of micromanaging.

A BIT OF PHILOSOPHY

That brings in another perspective, which is that of whether we use permaculture principles in our own lives. Is our human system one in which we're all participating and all have our preferred niche? Permaculture makes us ask some questions about our relationships with each other, particularly when we're thinking about 'care for people' and 'fair shares.'

It rejects the idea that we have to control nature, and it rejects the idea that destruction of the environment can ever be justified by monetary profits. The good of people and the good of the planet go together and are indivisible. But it can be uncomfortable when you start thinking about the wider world; if you have shares in oil companies, for instance, how does that sit with your permaculture ideals?

I'm certainly not telling you to sell your shares. There may be other ways to think about it; for instance, if you're taking your dividends and investing them in small energy-saving businesses, or if you're an activist shareholder, those are thoughtful responses to the situation. The key is not to shirk the thinking. When you take up permaculture, you become a responsible agent, responsible for looking after the Earth and its people, and that involves thinking about where you fit. Where is your niche in human polyculture?

Permaculture practitioners have often got involved in other movements. Some run social businesses or repair cafes; others teach natural building techniques or are involved in restoring old buildings using traditional, organic materials. Some are artists; some are storytellers; a number of permaculture teachers have worked hard to embrace social diversity, giving scholarships to POC students or helping give local women a voice.

Remember the concept that 'things happen on the edge' in an ecosystem? That's true of humans, too; the Beat Poets, the New Journalism, the Harlem Renaissance all came from the 'edge' of society. So maybe that's a place that a permaculturist can feel at home…

On the other hand, if what you want is simply to create a garden that is natural and organic, sure, you are still doing your bit for the planet. So I'll quit the philosophical musings and let's get on to looking at a further refinement of polycultures, the guilds.

Tips on growing your own polyculture:

- Keep it simple to start with. For many families, an annual vegetable bed is the easiest and most productive way to start.
- Start your annuals in spring by just scattering a mix of different fast-growing salad vegetables—lettuce, arugula, bok choi, radish, and spinach. That will give you ground cover and some great fresh food.
- Add some spinach and broccoli, directly grown or from seedlings.
- You don't have to wait for vegetables to mature; eat baby leeks, salad leaves
- Use 'stacking in time'—planting your winter-growing vegetables in August or September, then adding spring sowings as soon as the first frosts are past (or earlier, in a polytunnel).
- As you get more experienced, think about adding some perennials such as fruit bushes and creating polycultures around them.
- Welcome volunteers—self-seeding vegetables, flowers that arrive
- As you eat the early salads, you're naturally thinning out the bed. Let some of them go to seed and sow the next crop (the bees will also thank you for the flowers).
- Sow some carrots, beets and Swiss chard once the early salad has left enough space—they'll be ready in the fall.
- Try to mix slow and fast growers, so eating the fast plants leaves the slower growing ones some space.

such as 'mesclun,' or baby carrots, as soon as they're ready.
spontaneously, and friends who might do some work in return for fresh salad or apple pie…
- Be ready to improvise! Always fill spaces up with something new.
- If something works, do it again—if it doesn't (even if it's in this book!), then try something else.
- Remember to refresh the soil in the fall with a dressing of mulch.

EXERCISES

Build-Your-Own Polyculture

Take one of your favorite fruits or vegetables. You want to grow plenty of it. Now think about what needs it has, for instance, shade or sun, protection from pests, or plenty of nitrogen. It might be a herb, an annual like squash or melon, or a fruit tree. The only criterion is that you have to really, really like eating it.

Next, pick plants that you could grow with it to provide it with its needs. Pick about a dozen different possibilities and write their names on slips of card or paper. Now shuffle them around. Some of them might not work well together, even though they work well with your favorite. Which would you prefer to 'lose'?

Try to get down to three or four plants that create a mutually beneficial polyculture. You might even be able to build two different polycultures—if you've got room in the garden, why not try them both out and see which works best?

Planting patterns

We talked a bit about different planting patterns for intercropping. Design different ways that you could interplant two crops—in lines, diagonals, circles, zigzags, whatever. You might think about adapting your intercropping designs for keyhole beds. Draw them in two different colors so you can easily see how they fit.

Then step up and have a go with three different crops. You could have different relationships between them—one crop planted around the first, for instance, with the third crop making lines between the circles.

Just see how many different ways of planting you can find. Break your imagination free!

10

GARDEN GUILDS

WHAT IS A GUILD?

We've already talked about polyculture, companion plantings and intercropping. A *guild* is a special way that permaculturists refer to a known polyculture—a group of plants, microbes, and sometimes animals, creating a pattern of mutual support. Often, a guild is centered around one species, typically a large tree or shrub, and is named after it.

In a guild, all the elements come together to support one another. In a way, the guild combines both wildlife gardening and vegetable/fruit gardening and it benefits both humans and animals. It looks almost like a natural landscape. Depending on the size of your garden, it might look like a forest edge or a full-scale forest with its different canopy, shrub and ground cover layers.

A guild is a dynamic, self-sustaining polyculture. The idea is that it should provide its own needs for fertilizer and suppress weeds and pests, with minimal intervention from the gardener (the name 'guild' comes from the trade associations of the Middle Ages—companies which provided mutual support to their members and which regulated their own trades, instead of being regulated by the local prince or bishop).

There is some real science behind guilds, but not rocket science! Instead of selecting plants for color, prettiness, height, or foliage, we're selecting for function. In other words, we're looking at plants in terms of their job description. The plants are actors; they are dynamically doing something within their ecosystem.

You can design a guild for a full-scale food forest, but you can also design one for the corner of a suburban garden. The scale is different, but the principles are the same. Many large permaculture gardens which aren't full-scale farms are designed as a series of what you could call 'woodland islands,' focused on one, two, or three fruit trees.

You could have linear guilds. A narrow line (or even better, a curvy zigzag) of perennials between annual vegetables could establish this kind of guild. It can attract predator insects to make a bug bank, grow green mulch material like comfrey, clover and lucerne, and let you grow perennial herbs and vegetables like rosemary, thyme, chives, lemon balm, lovage, sorrel, perennial onions and oregano. If the perennials are tall, you could use them to shade one side of the vegetable bed and grow more shade-loving vegetables like lettuce on that side.

The three sisters—corn, beans, squash—are a classic guild. They nest and stack into each other, they help each other and all three yield food—in fact, the sisters have a higher yield together than any one of those foods planted separately in the same area of ground. Do note with this guild that it's important to plant the corn first and get it off to a good start before you plant the beans. When the beans shoot up, they need the corn stalk to be strong enough and high enough for them to start climbing.

But maybe you could even add to the sisters. As I mentioned, the Rocky Mountain bee plant (*Cleome serrulata*) was sometimes added to attract pollinators (it was also used for medicine and dyes). You could add chill peppers, sweet potatoes, or even comfrey. Some people have added tomatillos (*physalis*) and amaranth. Others have added sunflowers—which distract birds from eating the corn kernels—or bee balm (bergamot) as an insectary.

MAKING A GUILD

When you make a guild, you're aiming to create a healthy plant community that can recycle its own waste back into usable nutrients, cover the ground to prevent weeds growing, resist disease, control pests, harvest

How to build the Three Sisters Guild	Start by making a mound about four feet wide and a foot high
Corn	Plant six kernels of corn in a circle about two-foot diameter at the top of the mound—make sure the last frost has passed
Beans	When the corn is about a foot tall, plant four bean seeds around each stalk of corn
Squash	A week or so later, plant twelve squash seeds around the edge of the mound (six holes, with two seeds in each hole)

THREE SISTERS GUILD

and conserve water and attract insects for pollination. Each of those functions might be represented by a different plant, or a single plant might cover several of the functions.

Suppose you have a companion planting of carrots and onions. What plants could you add? Let's look for the right functions. Carrot and onion both have leaves that don't cast much shade, so weeds might take over the ground around them, or water could evaporate. You need ground cover. Carrot is prone to carrot fly and other pests. You need an insectary or a pest deterrent. Carrots also take quite a lot of nitrogen out of the soil, so you need a nitrogen fixer.

If you add lettuce, a ground cover, that will help keep the soil moist. You could add strong-scented rosemary to deter pests; it's perennial and can grow quite tall, so maybe put it at the center of the bed. Peas would give a nitrogen fix, but don't get on well with onions, so maybe put the peas and carrots to the other side of the rosemary from the onions; their nitrogen will still benefit the carrots. Their shade could help the lettuce, which doesn't like too much sun. Or maybe you could find a nitrogen fixer that will get on with your onions?

Usually, when describing guilds (and often when creating them), we start with the center. That's often a tree, but it doesn't have to be. The highest plant might be a bush or a big cherry tomato vine growing on a trellis. Tomatoes are hungry plants and don't give much to the other plants, so for that guild, you're going to want to look for other plants that give the tomato what it wants. For ground cover, for instance, you could grow nasturtiums, which are pretty, sprawl nicely on the ground and also, if you let them go to seed, produce 'capers' that you can pickle and that make just as nice a sauce tartare as the real thing. It's worth noting, however, that it's best to avoid planting these in newly manured ground because they can grow very large and leafy if over-fertilized, taking over places where you don't want them, such as paths in your garden.

Tomatoes need nitrogen, but they don't really like growing with beans, which would be our normal go-to for soil fertility. So a better choice would be a long-tap-root plant, a nutrient accumulator like comfrey, or borage—which also attracts pollinators and, usefully, repels tomato worms. You can then chop the leaves at the end of the season to mulch the soil and put some nutrients back for next year.

Another plant that gets rid of tomato pests is garlic, which will also ward off blight. You could also add asparagus, as a perennial and a good plant for spring, before the rest of the garden is producing fully. And you could add some basil because when you pick the tomatoes, you'll want to pick some basil too. Some say it even improves the taste of the tomatoes!

I always like to draw the guild out with the different relationships and functions in different colors. That lets me see if the whole guild would collapse if one of the plants failed and if there are any plants that aren't really contributing.

Ideal plants for a guild are those that can add two or three functions into the mix.

It can be a good idea to keep a spreadsheet or record book; however, you prefer to keep your information, with notes on each plant, including its preferences for sun or shade, moist or well-drained soil, its average growth

| **Guild considerations** | • Compatible water/moisture needs
• Compatible root systems (some deep, some shallow)
• Height of the different plants—vertical stacking |
|---|---|
| **Guild functions** | • Ground covers
• Soil feeders (mulch plants)
• Pest deterrence
• Insectary—attract pollinators
• Insectary—attract predator insects
• Feeders
• Climbers
• Rooters
• Nitrogen fixers
• Nutrient accumulators
• Supporters |

in terms of height and habit (e.g., spreading, or tall and thin), its root type (taproots, shallow or deep-rooted), its native area and its functions. Then if your guild doesn't work, you can go back and check why; for instance, maybe you have one plant that is excessively vigorous and a number of plants that are towards the extreme of their climate zone, which is why the first plant is taking over.

Remember, a guild ought to include all the major functions. If your guild doesn't cover all these bases, then it leaves you to do that part of the work. So, for instance, if you have a guild that doesn't have a good ground cover plant, you're going to end up doing the weeding and probably watering. In fact, the guild is a really good case for considering those two permaculture principles again; one element, many functions; one function, many elements.

If all you had was one tomato plant or one apple tree, you would have to do all the rest yourself—feeding the soil, picking off pests (or using pesticides), even pollinating the flowers yourself with a paint brush. Instead, the guild puts other plants and animals to work on your behalf.

APPLE TREE GUILD

Let's look at the apple tree guild—this is a real classic and in fact, it's a mini food forest. If you don't like apples, it works for other fruit trees too. Our center for this guild is the apple tree.

You could make a simple two-part guild here—apple tree and comfrey. The tree shades the comfrey; the comfrey provides mulch and nutrients and stops grass from competing with the apple tree.

But you can do better than that.

Apple trees don't like to have to work too hard for nutrients; they don't like having grass and weeds around their roots. So a ring of bulbs, whether that's daffodils (which look pretty in spring) or alliums (green onions for instance, or garlic) can surround the tree, roughly where the drip-line of the foliage is, and that will be the first line of defense against grasses and other perennial weeds that grow by reaching out with their root systems. They won't easily get past the bulbs. Garlic or garlic chives play two roles because they emit strong scents that many pest insects (and animals, like deer) don't like.

We need pollinator-attractors and remember when you're choosing them that you need your pollinator-attractors to be in flower at the same time as your apple tree is in bloom. You'll need to check the individual tree and when those plants flower in your particular area. So plants like bee balm, dill, or fennel, are all useful to have around. Fennel is particularly good because, like garlic, it has some anti-fungal properties, so it can help to protect your tree against apple scab (but remember fennel is allelopathic to most other plants and so will need to be grown away from them).

Thinking about the soil, we need plants that will help provide fertility. Comfrey is an obvious choice which can also be used as green manure. Dandelions and white clover also help accumulate nutrients; borage is another good choice.

Then let's just use a couple of other techniques to make sure our apple tree does well. First of all, let's sheet-mulch the whole area inside the drip line by covering it with cardboard, then a layer of compost, and watering it till it's nicely soaked. That should start off improving the soil. And then next, let's exercise a bit of self-discipline and *not walk there* at all, unless we are either pruning or harvesting so that we don't compact the soil, and then all the beneficial microbes have the oxygen they need to do their bit.

But this isn't the only apple tree guild you could make. Toby Hemenway suggests apple, redcurrant, fennel, mint, salvia and comfrey. The herbs are all insectary plants, everything is edible and the mint will be good ground cover.

Don't plant your guild too densely. The insect plants can be spaced out a bit. In a temperate climate, spread the guild out, so there is not too much shade. In a very hot climate, you may want to plant a little more densely so that the smaller plants are well shaded from the sun. Remember, leave the plants space to grow.

THE MAIN ELEMENTS OF A GUILD

Different permaculture teachers have slightly different ways of looking at the elements of a guild and sometimes use different terms for them. But the basis is fairly clear and centers on the functions of the plants.

Feeders are usually at the center. These are plants you're going to be eating—usually (not always) a fruit tree, sometimes a nut tree or bush. You might also think about other needs you have, such as basketry materials, building materials, or energy (firewood). They may also have the function of feeding wild animals such as birds.

Supporters are normally the feeders which support any climbers you have in the guild.

Coverers make sure the ground is covered and prevent grass and weeds from competing with the feeder plant for nutrients. Bulbs in a circle are good for the job, as they grow in the spring to deter the weeds but then die down in the summer when the tree needs more nutrients. In warmer climates, sweet potatoes, yams, cassava, squashes and gourds all help; in temperate climates, alfalfa, herbs and salad leaves are useful. Also, look for plants with big, fleshy leaves like rhubarb or squash.

Rooters are plants with deep root systems that dig and mine the soil, breaking it open and giving you the tilth you want. Depending on your climate, cassava, sweet potatoes and yams, or carrots and beets will fit the bill, or dandelions and comfrey.

Climbers can use the center plant or a trellis, or even the wall of a building for support. They improve the efficiency with which you're using the plot by stacking vertically for more food production. Kiwi fruit, passion fruit, climbing beans, cucumbers and even squashes can be trained upwards. Be careful because if you have a robust climber in a tree guild, it may try to take over the tree.

Protectors include insect and bird attractors like dill, comfrey and fennel, which lure pollinators and pest-eaters. They also include pungent-smelling plants, which will deter insect and mammal pests; think of lemongrass or garlic. Thorny plants can be used as barriers to deer and other grazers.

Insectary plants like lavender, bee balm, golden marguerite or fennel and coriander also fit in this element.

Soil feeders are plants which will help build fertile soil. There are three main types:

1. *Mulch plants* like comfrey, artichokes, rhubarb, or clover, which can be slashed and left as a mulch, which will encourage worms, microbes and fungi to work in the soil and will eventually be decomposed into humus.
2. *Nutrient accumulators* like chicory, dandelion, yarrow and plantain—these will eventually become redundant and die down as the nutrients recycle within the guild and are contained in the top layer of soil.
3. *Nitrogen-fixers* such as beans, peas, groundnuts and leguminous trees.

If you have these types of plants in the guild, then they will keep nutrients recycling in the guild so that you never need to add fertilizer. Although you remove nutrients through cropping, by using green manures and composting, you should be putting enough back into the system for it to come close to the efficiency of a closed system. If there is a small 'debt' at the end of the year, you can get some manure or straw to add to the system—but if you have chickens or ducks adding compost to the system, you may not even need that.

This is where permaculture goes beyond organic gardening. In organic gardening or farming, you have to keep composting and mulching all the time, but in permaculture, though you compost and mulch to get things started, you aim to build a system that will do the job for you once it's established and in which your intervention is needed less and less as time goes on.

The table shows the most commonly used plants for each function within a guild.

Remember that establishing a guild takes time. A guild also works through time, through the processes of succession, so some of the plants you'll need at the beginning may not stay all the way through and you may need to establish other plants to take up the running. Permaculture does take patience, sometimes!

FUNCTION-STACKING GUILDS

Think about each function that you would conventionally have to provide through your own labor. Now find a plant that does the job. Start with the most important function; for example, in a bird-attracting guild, we might begin with a couple of berry bushes that produce berries at different times of the year.

Guild plant functions		
Function	Description	Examples
Animal feeders	If you have domestic animals in your guild, you can provide food for them from your guild	Buckwheat, grasses, Siberian pea shrub
Attract pollinators	Attracting pollinating insects into the guild is necessary for the productivity of your fruit plants	Basil, bergamot, blazing star, coneflowers, cucumber, dill, evening primrose, indigo, joe pye weed, lobelia, mint, oregano, peas, rosemary, sneezeweed are all good plants to attract bees to your guild
Climbers	Climbers increase productivity by using up all the space available in the guild to produce food and other products. They can be grown up structures such as the house or trees and other plants. They are useful where land may be scarce. Uses the layers of the forest to its full potential. They can also be used as shade from the sun for certain species. Vigorous climbers can be trimmed back and used as mulch	(Perennial species) choko, passionfruit, kiwifruit, hops, sweet potato, scarlet runner beans and jasmine, (annual species) cucumber, climbing beans and climbing peas
Competition barriers	Competition barriers can be planted around other plants to prevent competition from grasses, weeds and other invasive plants	Daffodils, comfrey and perennial alliums (garlic chives, wild leek or Egyptian onions), oats, buckwheat, gooseberry
Rooters	Deep rooted plants such as trees reach far into the soil and bring minerals up to the surface. They break open the soil, leaving it soft and allowing for air & water to be easily absorbed into the earth	Trees, root crops (potatoes, sweet potatoes, yams, carrot, beets, daikon radish, parsnip), dandelion, comfrey, chicory, parsley, fennel, dill, queen anne's lace, avocado, mallow, burdock
Erosion control	These plants hold the soil in place	Buckthorn, bamboo, sumac
Flood management	These plants can stand submersion in water	Annual ryegrass, fountain grass
Food for us	Maximise the nutritional benefits you will be getting from your guild. Plant a diversity of foods. Include foods from all 6 food groups so that you are fed all through the year	Vegetables, staples, legumes & fruits, fats & oils, animals. You will probably want to focus on perennials such as certain spinach plant, chicory, globe artichoke, Jerusalem artichoke

Guild plant functions		
Function	Description	Examples
Green manures	Green manures are a type of mulch which are allowed to grow before the point of flowering then slashed to provide plant material which can be used as mulch or dug in to provide fertility to the soil	Clovers, barley, beets, buckwheat, mustard, rocket, millet, purslane, turnip, wheat, basil, dill
Ground covers	Another type of mulch, ground covers form a living mulch which protects the soil by growing a barrier of plant material between the soil and the elements	Pumpkin, squash, strawberries, clover, sweet potato, creeping thyme, creeping rosemary, mint, alfalfa, vetch, oregano, marjoram, sedum, comfrey, yarrow, chamomile, catmint, ladys mantle, chives, violet, nasturtium, periwinkle, dewberry, lungwort, sorrel, ajuga, sweet cicely, dogwood Lettuce and other salad leaves
Nitrogen fixers	One of the main nutrients plants use for food is nitrogen. Plant nitrogen fixers to put nutrients into the soil; this also helps feed the plants growing nearby	Legumes (broad beans, dwarf and climbing peas, groundnuts, kudzu, leguminous trees, clover, alfalfa, cowpeas, lupines, vetches, soybean, peanut) Lupin, clover, ceanothus, kaka beak, dyers greenweed Trees and shrubs (tree Lucerne, sea buckthorn, Siberian pea shrub, Elaeagnus, black locust) Robinia (black locust tree, golden rain, alder)
Feed the soil	Return all organic matter back to the soil to provide it with nutrients	Leaves, trimmings, kitchen scraps, decaying matter, compost, manure, urine
Mulch makers	Mulchers produce large amounts of soft leaves which can be cut down and left on the ground (chop and drop) to break down, forming a mulch and releasing nutrients	Comfrey, Jerusalem artichokes, rhubarb, nasturtium, plantain, dock, deciduous fruit trees, beans, burdock, mallow

Guild plant functions

Function	Description	Examples
Nutrient accumulators	These plants use deep roots to access nutrients which are out of reach for plants with more shallow roots. They accumulate nutrients in their leaves which can be cut as mulch to help other plants	Borage, stinging nettle, comfrey, broad-leafed dock, horsetail, parsley, plantains, yarrow, dandelion, chicory, tansy, nettles, chives, valerian, chamomile, calendula, rhubarb, sorrel
Pest control	To keep harmful pests under control, you can attract beneficial insects to your guild by planting certain plants that they're attracted to. You can attract pollinators, predators and parasites	Bergamot, buckwheat, coriander, alfalfa, goldenrod, marigold, asteraceae family (daisy family), umbelliferae family (carrot family), alyssum, calendula, chicory, cosmos, gypsophilia, lavender, lupins, nasturtium, queen anne's lace, yarrow, bee balm
Pest repellents	You will want to protect your guild from damaging pests and this can be done with certain plants that confuse pests with their aromatic smells, making it more difficult for pests to identify their target plants	Pest confusers – basil, chamomile, coriander, dill, fennel, garlic chives, horehound, lavender, marigold, mustard, nasturtiums, onions, lemongrass, pennyroyal, rue, tansy, rosemary, wormwood, mint, horseradish and yarrow
Shelter	Certain plants can provide shelter for birds but also protect trees from wind or sun damage on species such as oak, macadamia or paw paw	Jerusalem artichoke, cana lily (protect trees), hedges (shelter for birds)
Soil protection	Groundcovers protect the soil from sun, help hold moisture and prevent weeds from spreading	Sweet potatoes, vines, pumpkin, cucumber
Supporters	These support the climbers. Some climbers can be aggressive and bring down a supporter so think about this when choosing where to put a climber	Trees, shrubs, bushes, stalks, outdoor structures (house, tables etc). Plants that act as supporters include Jerusalem artichoke, sweetcorn, dent corn, fruit and nut trees
Wild animal feeders	You can incorporate beneficial plants into the guild to provide food for birds	Edible trees, plants and bushes such as nuts, berries

Build the other functions around that function. You need to fertilize the soil; which plants will you use for that? You'll need to attract pollinating insects; which are the right plants here? You'll need to deter pests; which plants can do that? Cross-check that none of the plants you choose are antagonists of others.

It's not rocket science. The single most important factor is determining what, for you, is the most important function. It may be edible food; it may be attracting birds (many people put guilds like this near their living rooms or kitchen windows so they can see the birdlife); it might be providing for your honey bees.

In fact, there are a number of guilds that are well known in permaculture circles and that I will outline in a little while, so you can take those as your basic recipes.

MORE ADVANCED TECHNIQUES

You can start with guilds like the apple tree guild as your 'cookbook'. But once you have a bit more experience, you may be better off designing your own guilds for your garden's specific climate and microclimates.

Observation

Wander through local forests and woods or hedgerows and see which plants grow well together. Are there certain combinations that you always find together? Where I live, for instance, elderberry and blackberry always seem to be in the same hedges with crab apple trees. The elderberry grows fast and straight and the blackberry uses it as a support. So that's something I've incorporated in my garden plans.

Research

There are plenty of good ways to search for plant communities in your area. The USDA forest service, plant associations, and state and government ecology websites all have information on native plants. The United States Botanic Garden at www.usbg.gov has lists of recommended native plants and information on urban agriculture. Sometimes tourist offices have resources on nature sites and local plants and gardens and these can be very specific. Local museums rarely have great displays on local plants (though they may on traditional crafts that used local plants), they are often a good place to ask questions, as they may know the right people to ask, or have a research department that can help.

Once you have your list, do some more research. What functions do these plants provide? Were they traditionally used in particular ways? For instance, I found out that walnut husks used to be used by local woodworkers and basket makers as the base for a black dye. I even got the recipe from one of the older people in our town.

If you know other organic or permaculture gardeners locally, ask them for their suggestions too. And then, once you've found out about the plants, try them out!

Orientation

This little finesse can help you get the most out of your guild. If you have a 'forest edge'-style guild, put the taller elements to the north (assuming you are in the northern hemisphere) so that all the elements of your guild get sun during the day—*except* in a desert environment, where you would turn it around so that the whole guild gets shaded all day. Or, if your guild is around a tree, plant the shade-loving plants on the north and the sun-loving plants to the south, particularly the south-west for the real sun-lovers. South-east, which gets the sun in the cooler morning hours, is best for partial-shade lovers. As the sun changes position during the day, your plants will all get the amount of sun they like.

Habitat nooks

A nook is slightly different from a guild. When you build a nook, you're trying to create a small niche of habitat where lizards, frogs, snakes, or small birds will find an attractive home. The slow worm (not a snake but a legless lizard) in particular loves eating slugs and can be a great gardener's friend. Frogs will eat mosquitoes and other insects. Toads eat both slugs and mozzies; they like to hang out in little crevices. A friend of mine has a toad that lives in an upside-down flower pot with a crack in the side and a bit of moss on top to keep it cool and damp. Since he adopted the toad, he doesn't have a slug problem anymore.

If you've ever had an apple with a worm in it, that's the work of the codling moth. Bats eat those moths in their thousands—that is, thousands every night! Put up a bat-nesting box somewhere high where the bats won't get disturbed—the top of the house gable or the side of a large tree would work. Old farm buildings with unused attics often house a bat colony—if you have an old barn, check the floor for bat droppings.

We've already talked about attracting birds with feeders and nesting boxes, but slow worms and lizards prefer a refuge that's well hidden. If you lay a bit of old carpet down somewhere, slow worms will hide underneath it; snakes like wood piles, particularly if the bottom is a bit rotten (that's a good reason not to stick your hand in a woodpile without looking first, by the way). The more micro-habitats you can make, the more likely you are to attract a range of predator animals that will keep your pests in check.

A slightly different use of nooks is using basins filled with mulch as nooks to harvest rainwater. Over time, this should reduce the need for irrigation in your garden.

COMFREY, EVERYONE'S FAVORITE

If you talk to any organic gardener or permaculture gardener, sooner or later, you'll find out about comfrey—the miracle plant!

It doesn't look particularly special, with floppy leaves and little, pinkish flowers, though it's not unattractive. But it's a really functional plant. Since it grows in zones 3 to 9 and is very tolerant of different growing conditions, it can be recommended for most gardens. You can plant it right next to a spillway and it will thrive, but it will grow in the sun too.

Comfrey attracts pollinators and its big leaves create a mini-habitat for beneficial insects. Like borage, its leaves are slightly furry or prickly (depending on the type of comfrey you have), so they deter snails and slugs. It has deep roots, so it's a nutrient accumulator, reaching down into the soil for nutrients. The leaves can be chopped up to create a nutrient-rich mulch or for composting—it's a good compost activator. It also creates a huge amount of biomass in a little space, so it's a great green manure crop.

You can dry comfrey leaves and powder them to make a growth accelerator, which can be mixed into the soil before planting in spring. A few comfrey leaves buried under newly planted trees, bushes or herbs will gradually release a nutrient boost. And true comfrey (*Symphytum originale*, not *x uplandicum*/Russian/Bocking) self-seeds prolifically, so if you have a problem area, it will sort it out over a couple of years. Russian comfrey won't self-seed, but you can propagate the clumps by division (which might be a safer option if you have a small garden).

Comfrey is even medicinal; it makes a great poultice for sprains and aching muscles. It fails on a single function; it's not edible. In fact, it's toxic.

SAMPLE GUILDS

These are only samples. You can build up your own guilds—and you will, over time—but these are good ones to get started, as they have been used by other permaculture gardeners who have had success with them.

Apple guilds

A really simple guild: blackberry and apple; the bramble protects the tree and both are edible (a standard pie combination).

Getting more complex: apple tree, garlic (grass suppressing, pest-deterrent, anti-fungal), nasturtium (pest-repellent, mulch-maker, attracts beneficial insects, medicinal), onions (grass suppressing, pest-repellent) and daffodils (grass suppressant) around the base of the tree.

Add comfrey (mulch, ground cover, insectary, nutrient-accumulator), yarrow (nutrient-accumulator, attracts beneficial insects), fennel (attracts pollinators and other beneficial insects, anti-fungal), chicory (*Cichorium intybus*, insectary and nutrient accumulator), chives (protector, pest repellent, edible), sweet cicely (*Myrrhis odorata*; insectary and aromatic), clover (insectary, ground cover, nitrogen-fixer).

Walnut guild

Walnut (timber, nuts to eat and make oil, husks for dye); walnut husks are sometimes used as a herbicide mulch. However, a trial by Sustainable Agriculture Research and Education (SARE) showed that walnut husks soaked in water were ineffective and walnut husks soaked in vinegar were only as effective as vinegar used on its own as a pasture herbicide.

Redcurrants, comfrey, daffodils around the drip-line; hemerocallis (daylily) and hostas for ground cover.

Remember that walnut's allelopathic chemical, juglone, will deter most plants from an area about one-and-a-half times bigger than the tree's drip zone. In a lot of gardens and old farmyards, I see the walnut tree put in a corner, all alone. That's why. This allelopathic effect is where the mulberry guild comes in because mulberry is a good barrier tree between walnut and other trees; sour cherry or black cherry is another. The barrier trees and their guilds can help ensure the walnut's influence doesn't spread too far.

Mulberry guild

Mulberry (*Morus nigra*, edible), kiwi fruit vine (edible), comfrey, licorice (nitrogen-fixer and insectary), sorrel (edible, nutrient-accumulator), sweet cicely (ground cover).

Or; mulberry, comfrey, dill (insectary), nasturtiums, clover (ground cover), marigolds (potential pest-repellent).

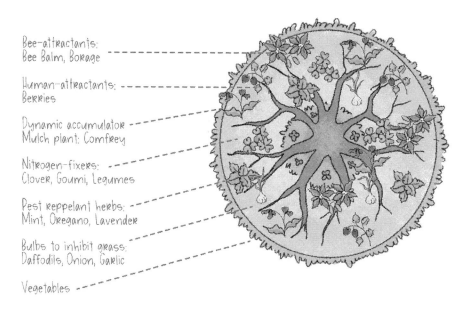

Bee-attractants:
Bee Balm, Borage

Human-attractants:
Berries

Dynamic accumulator
Mulch plant: Comfrey

Nitrogen-fixers:
Clover, Goumi, Legumes

Pest reppelant herbs:
Mint, Oregano, Lavender

Bulbs to inhibit grass:
Daffodils, Onion, Garlic

Vegetables

APPLE TREE GUILD DESIGN

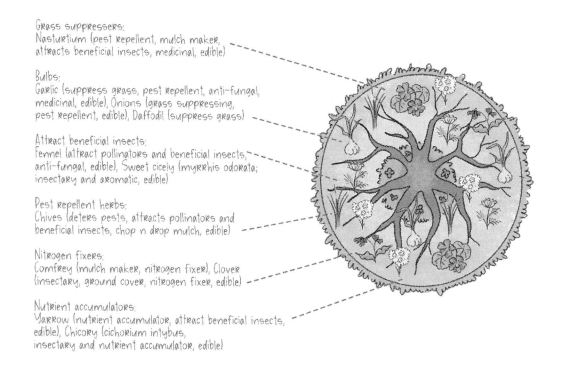

Grass suppressers:
Nasturtium (pest repellent, mulch maker, attracts beneficial insects, medicinal, edible)

Bulbs:
Garlic (suppress grass, pest repellent, anti-fungal, medicinal, edible), Onions (grass suppressing, pest repellent, edible), Daffodil (suppress grass)

Attract beneficial insects:
Fennel (attract pollinators and beneficial insects, anti-fungal, edible), Sweet cicely (myrrhis odorata; insectary and aromatic, edible)

Pest repellent herbs:
Chives (deters pests, attracts pollinators and beneficial insects, chop n drop mulch, edible)

Nitrogen fixers:
Comfrey (mulch maker, nitrogen fixer), Clover (insectary, ground cover, nitrogen fixer, edible)

Nutrient accumulators:
Yarrow (nutrient accumulator, attract beneficial insects, edible), Chicory (cichorium intybus, insectary and nutrient accumulator, edible)

APPLE TREE GUILD DESIGN

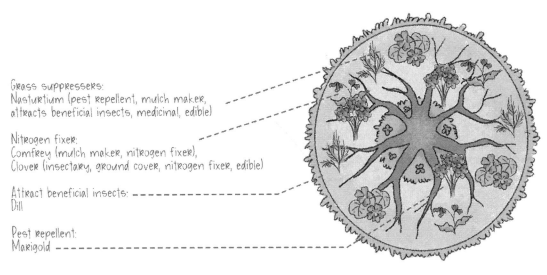

Grass suppressers:
Nasturtium (pest repellent, mulch maker, attracts beneficial insects, medicinal, edible)

Nitrogen fixer:
Comfrey (mulch maker, nitrogen fixer),
Clover (insectary, ground cover, nitrogen fixer, edible)

Attract beneficial insects:
Dill

Pest repellent:
Marigold

MULBERRY TREE GUILD DESIGN

Hazel tree guild

Two hazels (needed for pollination; get the type that performs best locally), serviceberries or elderberries (underlayer, edible), redcurrants or gooseberries (edible), pole beans, which can climb in the hazels (nitrogen fixers and edible), lovage; comfrey, horseradish can occupy the outer edges and you can use mints, low-growing roses, or clover as ground cover.

You will need to coppice the hazel after about eight years, when yields start to fall, by chopping through the trunk close to the ground. It's drastic, but you will see a load of new trunks springing from the outside in the new year. Some forest coppice stools near us are impressive, reaching nearly twenty feet across, as the hazel has kept sprouting outwards again and again. They must be nearly a hundred years old.

Or you can grow this guild as a hedge or windbreak: it's good for screening the borders of your property.

Peach tree guild

Peach tree; comfrey; garlic on the outside (pest-repellent), strawberries (ground cover, edible), red clover (nitrogen-fixer, ground cover), cosmos (insectary), chicory (accumulator, insectary).

Or, since peach trees don't cast as much shade as other trees, you can have more sun-loving plants in the guild; peach, rosemary, marigolds, cucumber, or squash, zinnias, arugula/rocket.

Pear tree guild

Pear; mint (edible, ground cover), columbine (insectary), borage (insectary), clover (nitrogen-fixer), scallions, day-lily (insectary), calendula.

This is a great guild that comes from the US west coast. The scallions, planted as a ring, keep grass and other perennial weeds out.

Fig guild

Fig tree, comfrey, squash, seaberry (nitrogen fixer), canna lily on the southern side for extra sun.

Wet meadow guild

This guild isn't based on a particular plant but on the question of what you do if you have a wet area that just doesn't drain.

Centerpiece: shrub willow (e.g., *Salix discolor*), which loves the wet and will produce materials for basketry and attract birds. On the outskirts of the wet area, three persimmons (kaki, *Diospyros virginia*, edible). Highbush cranberry (shade tolerant, edible) and dogwood (*Cornus canadensis*, wet- and shade-tolerant ground cover) can then be added to the mix.

You could also grow chufa (*Cyperus esculentus*), groundnut (*Apios americana*), marsh woundwort (*Stachys palustris*) and flag iris in this guild.

Jajarkot tree guilds

These guilds come from the Nepalese permaculture project at Jajarkot.

Autumn olive (*Eleagnus umbellata*) is used for nitrogen-fixing. Apple and pear or plums, with redcurrants, other ribes family and raspberries just outside the drip-line. Spring bulbs are used to sequester nutrients; trout lily captures phosphorus. Other plants are used for ground cover and to attract insects; sweet cicely, dill; parsley, strawberries, white clover (in the sun) and wild ginger (*Asarum canadense*).

Bee guild

This guild, unusually, doesn't focus on an edible plant but on another edible product, honey. Highly recommended for beekeepers.

Linden tree (*Tilia cordata*, lime—good carving wood, flowers which attract bees), *Rosa rugosa* or *villosa* (produce hips, petals for tea, attract bees), lovage (taproot, nutrient-accumulator, edible leaf and seeds), *echinacea* (coneflower), comfrey, salvia and/or mints as ground cover, dill, fennel and caraway will self-sow (aromatics).

Beach plum tree guild

Beach plum (*Prunus maritima*); ramps (edible, protectors), camas (insectary and edible), dwarf coreopsis (ground cover, insectary). However, this guild doesn't yet have a nitrogen-fixer. Can you think of one to add?

Paw-paw guild

Paw-paw (*Asimina triloba*, edible), ramps, hog peanuts (edible, nitrogen-fixer)

Tropical guilds

Palms, as the canopy layer, then leguminous trees, fruit and nut trees, fruit and nut shrubs, coffee; pineapple, ginger, cassava, sweet potato (Bill Mollison used this one). Mango, banana, papaya, passion fruit, cassava, moringa, katuk.

Mango, carob or avocado, underlayer of papaya, guava, kudzu and groundnut (nitrogen-fixers; you may not need these if you used carob, which is itself a nitrogen-fixer), amaranth and moringa (dynamic accumulators), gourds and sweet potatoes (ground cover), lemongrass (pest-control).

THE DOWNSIDES OF GUILDS

There are a few downsides to guilds, so you ought to think about those before you establish your guilds. First, they do take up a lot of space, despite the fact that you can use vertical stacking and interplanting. More seriously, they are mostly slower to establish than an annual bed. Getting a guild to become self-sustaining can take five years or more.

Because you have so many plants all interrelated, it can also be difficult to work out the source of problems. In a monoculture, it's much easier to establish the exact reason that a particular plant is not doing well. Even in the annual garden, usually it's a bit easier because when you plant in rows, you only have relationships with the plants on either side.

And guilds can be pretty site-specific. The apple guild, walnut guild and variations of the three or four sisters should work in most of North America and Europe, but subtropical and tropical gardeners will find their needs are a bit different. And some plants aren't appropriate for guilds, so they will have to be planted separately.

Annual polycultures

Jajarkot, in Nepal, has developed an annual polyculture using the same concept as guilds. It starts in spring with greens providing ground cover, intercropped with slower maturing plants like beans and alliums. Although you can sow these vegetables every year, many are self-seeding if you allow a certain amount of your crop to bolt and go to seed (you can then pull the stalks out of the ground and lay them down for mulch/compost).

Ground cover includes mustard greens, mizuna, tatsoi, garden cress, arugula. Add some salad leaves, lettuces, chard, radishes. Add some carrots. Then sow herbs, fennel, dill and coriander. Finally, add the beans to the mix, poking fava beans, bush peas and so on into the soil about a foot apart. Then add the alliums—garlic, garlic chives, leeks, scallions. All of that gets sown pretty much at once and the ground cover will start growing really fast so that within a couple of weeks, you can already pull out some of the young plants to put in a salad.

After a month or so, you'll have a few gaps; this is when the cabbage, cauliflower, broccoli and other late-maturing, big plants can go in—best to transplant seedlings for these rather than sow direct. By early summer, basil and bush beans can be sown.

This polyculture can yield food for six to eight months of the year and by far the greatest part of the work is done with the early sowing, which should be at the last frost date in your area. After that, it's just adding a few extra plants and harvesting, which frankly most of us don't really consider work, as we're going to get to eat what we harvest in the next half an hour or so!

EXERCISES

Guilding the lily

Take one plant that you like. It can be edible, a tree, or a flower; I used *Hemerocallis fulva*, the orange day lily.

Does it already fit in one of the guilds we mentioned? Draw a chart showing its relationships with the other plants in the guild. How strong is the network of relationships?

If we haven't mentioned it, then find out about its qualities and see if it would fit into one of the guilds as a replacement plant. Or you might even have to design a guild around it, as you did with the polyculture exercise.

A walk in the park

Go for a walk in the park. Take a good look at what's there. How much biodiversity exists? It may differ between different areas.

For instance, in my local park, there's almost zero biodiversity in the very tightly controlled lawn—not even any weeds (they must be using pesticides). The same is true for the flower beds, full of annual flowers that were grown in greenhouses and transplanted.

But there's a wooded area where people walk their dogs that has much more biodiversity. There are at least five different types of tree and there's lots of undergrowth. I've noticed a few creepers and vines using the trees as support.

Take your sketchbook or notebook with you and jot down your observations. You may actually be lucky enough to see a few natural guilds getting established.

11

THE FOOD FOREST

The food forest is taking permaculture to the next level. You're building on the same concepts that we used in putting together fruit tree guilds, but instead of having an orchard nicely spaced out, you're putting the plants closer together and creating a multi-layer food forest. This food forest should maintain itself and provide you with food and other materials.

It won't all be thick forest. There will be clearings and there will also be some areas where the taller trees give way to shorter and sparser plantings. A food forest isn't a monoculture like a plantation, so it needs to have some niches within it for plants that prefer more open conditions too.

In England in the middle ages, the word 'forest' used to mean a parkland that included open spaces as well as woodland. So your particular food forest might feel more like that and less like a lush tropical rain forest. You could call it a forest garden if that feels more like the effect you're going for.

You'll need room for it… and it certainly won't start at your back door, more likely zone 3 or 4. But you can use food forest principles to build a … food wood? Food copse? Food thicket? So don't be put off by the word 'forest' and think you need a huge ranch to create one. If you can make an edible hedge, then what you've actually done is built a one-sided food forest!

As we saw when we talked about natural systems, natural succession starts with annual plants and almost always ends with a forest. It doesn't work that way in permafrost or in salt marshes, but it does almost everywhere else. Some wetlands gradually get filled in, first of all by rushes and reeds, then by thirsty trees like willow and alder, whose roots trap soil and stabilize the banks and then, in a decade or so, you have soil where you once had water.

That's happened to the Norfolk Broads in the UK. Lakes were formed as water invaded the depressions that had been made by peat digging. They have gradually shrunk over the last fifty years as alders have established around the margins, forming carrs—waterlogged woods that become a succession stage between marsh and forest. Remember the power of edges? This is a great example.

As a climax community, the forest is a stable ecosystem. It doesn't have any more succession to do, so it should be self-sustaining. The plants look after themselves: they don't need anyone to fertilize them because the forest delivers a leaf mulch every fall and the deep-rooted plants are always bringing nutrients up to the surface. They don't need anyone to seed them; they do that themselves, too. They don't need pest distracters or repellents because those plants naturally grow there, and there's no need for you to use pesticides.

The only difference between a wild forest and your forest is that you have designed your forest to include plants that will provide you with food. Otherwise, it should be the same—self-maintaining, self-replicating and self-sufficient.

We're modeling the system with productive species. Swap out the oak and ash trees and swap in fruit and nut trees. Swap out some of the ferns and replace them with perennial herbs. Swap out that thorny bramble patch and… no, you don't need to, as long as you like blackberries! And the great thing is that once your food forest is planted and growing, it should require the minimum effort from you.

Food forests go back thousands of years. We might think 'hunter-gatherers' just wandered around, picking what they could. In fact, they were much smarter than that. They worked out that they could encourage the plants they wanted to grow where they wanted them.

There's evidence that well over a tenth of the Amazon basin had been 'engineered' by forest gardening. Geoff Lawton found a food forest that he reckons was two thousand years old in Morocco—a date palm oasis, but with an understory that included huge olive trees, figs, carobs, pomegranates, large citrus trees, mulberry, quince, grapes and bananas. It's still being farmed. In Vietnam, he found a family working the food forest that they started three hundred years ago.

A food forest will evolve over the years. You need a bit of patience! Remember, the forest is the climax system of natural succession and can take fifty years to get established and even longer to become fully mature; we can and do accelerate that with permaculture, but it still takes a while. As the forest evolves, some plants will give up and others will take their place; the nitrogen fixers, which were essential right at the beginning, become a bit less important as the soil improves. Some of the plants that grew around the drip-line of young trees will find, as the tree grows, they don't like that much shade, so you may find they die off, or they scatter seed further outwards, even to the edge of the forest. So there is a long-term development, which isn't the case with a conventional garden, which only goes through the yearly cycle of growth and dying off.

So what does a food forest look like?

Like any forest, the first thing you'd notice about it is that it's multi-layered. There's a canopy layer high above your head; then there's a hedge and shrub layer perhaps with some smaller trees, then there are tall undergrowth plants and climbing plants that use the trees as their supports and then ground cover plants and flowers. There might be a clearing with some vegetables growing in it that forms another layer.

You can have as many as eight separate layers. I'll give some examples for forests in different climates. Depending on the space you have and the climate you're in, you may not be able to incorporate all the layers. For a mid-size garden, think in terms of three: trees, shrubs and ground plants. The larger the garden, the more layers you can accommodate.

1. Canopy—the tallest trees, thirty feet high or more. Oak, palm tree, mango, jackfruit.
2. Sub-canopy—shorter trees. If you have a smaller garden, this will probably be your top layer and it will include a lot of fruit trees. Apple, pear, feijoa, guava, citrus, pecan and coffee are some examples. This can be the most productive layer and the

most money-making for commercial food forests.

3. Shrub layer—perennials up to ten feet high, including smaller nut trees, blueberries, raspberries, elderberry, witch hazel… or bananas and tamarillo.

4. Herbaceous layer—many of these are perennial but don't have woody stems, so they die back to the roots in winter. Examples of these would be asparagus, kale, rhubarb and sorrel. You'll also find many annual herbs here (some people think bananas belong here because although they are tall, they die back to the rhizome after fruiting).

5. Ground cover layer—these plants are more shade-tolerant; examples could be mints, wild strawberries, creeping rosemary and thyme, oregano, nasturtium, pinto peanut, or Brazilian spinach.

6. Climbers and vines—these plants connect up the other layers. You could include grapes, beans, tomatoes, passion fruit, kiwi.

7. Underground layer or root yield—root crops like alliums, Jerusalem artichokes, ginger, taro and cassava. Sometimes, these plants are also good ground cover; for instance,

sweet potatoes are effective in both categories.

8. Mycelial layer—fungi; the soil in most forests is full of mycelia—long, stringy fungus 'roots' which help move nutrients and water around under the ground. Oyster and shiitake mushrooms are easy to add to your forest by buying an inoculated log or using your own hardwood logs plugged with inoculated dowels. In fact, it's the fungi rather than bacteria that tend to drive soil fertility in forests.

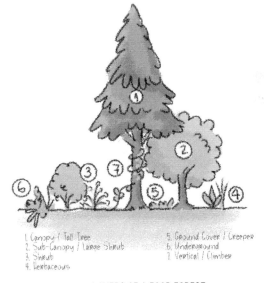

1. Canopy / Tall Tree
2. Sub-Canopy / Large Shrub
3. Shrub
4. Herbaceous
5. Ground Cover / Creeper
6. Underground
7. Vertical / Climber

LAYERS OF A FOOD FOREST

LAYERS OF A FOOD FOREST

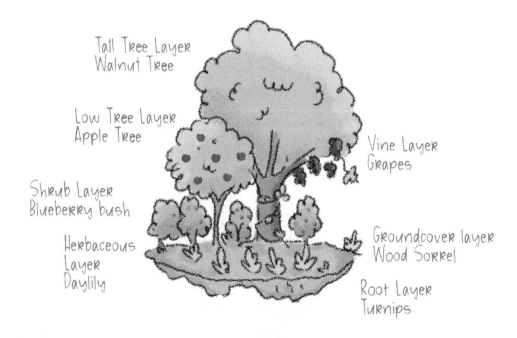

LAYERS OF A FOOD GARDEN

Remember that you may also have a pond or stream or even a river frontage which gives you a ninth area—an aquatic layer—together with the riparian edge, where many marsh plants can grow. This may also create cooler, damper microclimate niches.

The trees dominate, but they don't crowd out the other plants. And don't forget that with clearings, or the space where your zones 1 and 2 meet the forest, you've got an edge—a super-abundant space where both ecosystems can share resources. Dappled shade is a natural habitat for many perennial fruits and vegetables.

You can also think about different layouts. For instance, you could have alley cropping—leaving alleys between the higher trees for the fruit bushes and vegetables. Some Kenyan farmers have cabbage patches between palm trees. You could have an orchard-style layout of fairly regularly spaced trees, but unlike in most orchards, these are underplanted with shrubs and ground cover. You could have a more mature forest garden, which is irregular with some clearings and plenty of undergrowth, like mid- to late-succession woodland. Or, if you have an extensive area, you might actually have a closed-canopy forest, the forest at its most impressive and most mature.

If you have a large enough area, you could also include two or more types—for instance, an underplanted orchard in zone 3, a true forest garden in zone 4, leading to a closed-canopy food forest at the edge of your wilderness zone.

CREATING THE FOREST GARDEN—DESIGN

First, think about your aims and objectives. Edible yields are a top goal for most people, but you may also have other priorities. For instance, you may be interested in medicinal plants. For Native Americans, the forest is a source of medicines and that's true of many other forest-based cultures. You may want to generate fibers and craft materials; you could coppice hazel or willow for baskets or grow bamboo for basketwork. If you're living off-grid, you probably want a source of firewood for heating and cooking.

You might also think about it as a relaxation space, a space for festivals or parties, or an educational project. You might want to give small local wildlife an area that's more friendly to them than a conventional garden. Or, if you have children, you may think about it as a great place for them to explore and use their imaginations—perhaps a treehouse and a den are on the list of garden features.

As always, with permaculture, the next thing you need to do is to look at the site and do a zone/sector analysis (if you need reminding, go back to Chapter 2 and check the design principles). If you already have your site map, you can start drawing in the elements you need.

If you start from scratch with a flat lawn, for instance, the first thing you're going to need is a boundary—a shelterbelt to stop winds damaging your trees or shaking the blossom off your fruit trees. A hedge can cover an area about eight times its height; it needs to be planted quite densely. Or, if you have a slope, you might think about swales before you think about planting to make sure you've got the water well controlled.

There are a couple of other basics you should consider; any local law, such as limits on the height of plants, particularly overhanging boundaries, and whether there's terrain that needs stabilizing, such as a steep slope or a riverbank. These may dictate what you can do and where, as well as your priorities.

Then you can think about getting your canopy designed. If you have a large area and particularly if the soil is degraded, think about big trees that can produce nutrients and shelter for the other plants. Black locust, for instance, is a good nitrogen-fixer. Draw them in on your map, with the tree diameter showing the mature size of the tree (not what it measures now). Trees should be placed from a half to a quarter their diameter apart. So two four-meter diameter trees could be set one to two meters apart, depending on how dense you want that area of the forest (if the trees are different sizes, take the *average* of the two diameters).

Because a forest garden is three-dimensional, it might not be a bad idea to draw the elevation, as well as the plan, so that you can see how the plants will fit when they reach their mature heights.

Rootstocks

Most fruit trees are not commercially grown from seed. They are grafted onto a given rootstock. The graft determines the type of fruit (Golden Delicious, Granny Smith, Gala), but the rootstock determines the size of the tree. You can get the same variety of apple tree on different rootstocks. One might grow to thirty feet high, another to just six feet high, depending on the rootstock that was used. If you have a smaller space, you might want to take a semi-standard rather than standard rootstock, for instance.

Don't put too many trees in; they will grow to fill the space! Make sure you know exactly how large the mature trees will be, both in height and in spread. Leave some clearings—you don't want a forest that's too dense. You need to have room for shade, part-shade, and some sun-loving plants, so you should design in some edges.

Before you go on, look back and check the sectors on your design; check where the wind and the sunlight are coming from. If you have left gaps between the trees, how much sun will the plants there get? Have you set up effective windbreaks? Remember, with a food forest, the real work is in the planning. If you plan a conventional garden badly, you can always change things around with a bit more digging, a bit more spraying, whatever. If you plan a food forest badly, it can take some time to undo your mistakes.

Once you have finalized the tree cover, then you are ready to design the underlayer. For instance, you may be using fruit bushes like the Ribes family (redcurrants, gooseberries). You could increase your yield of tree fruits by adding apples and pears on dwarfing rootstocks that will keep them small if you want.

Try to add some nitrogen fixers in the shrub layer, such as eleagnus or Siberian pea tree—these are really important early on, though as the soil becomes more fertile, they will become less important and some of them can be replaced by edible species, such as Sea buckthorn.

Finally, you're ready to design the ground layer. This has several functions to fulfill. It can include plenty of edibles, such as lovage, Turkish rocket, wild strawberries and perennial onions. It also has an important task, to prevent soil erosion and provide living mulch to keep the soil moist, as well as to prevent plants you don't want taking over (stinging nettles, for instance).

A lot of permaculture gardeners like to include aromatic plants such as lemon balm (Melissa) and mints. You'll also want some nitrogen-fixers like vetch and some deep-rooted sorrel and comfrey as nutrient accumulators.

You might decide to grow fungi. The easiest way to do this is on spore-inoculated logs, which you can buy from many organic gardening suppliers. Then you know what you're eating is not a toxic wild mushroom! A shiitake log can keep producing fungi for several years. You may recall we mentioned how fungi help the forest floor when talking about hugelkultur in an earlier chapter. This is a similar process.

Don't forget to design paths that are wide enough for your needs. You need to be able to walk down the path carrying a basket and you might want it to be wide enough for a wheelbarrow.

Summary tips for designing your food forest

- Remember to look at the site first. We always design to take the best advantage of the site.
- Design in layers right from the start. Decide on your top layer depending on the size of your garden.
- Start setting out the top layer (your largest trees).
- Draw the tree on the plan at its mature size, not the size it is now, to leave room for growth!
- Design the underlying layers around the trees.
- Don't plant too densely. Allow for light and sun to get into the forest.
- You'll need a lot of nitrogen-fixers and nutrient accumulators early on. Later, other plants may take up the running.
- Don't forget good access plans.
- Don't plant seeds too close to one another.
- For small backyards, plant more verticals.
- Ensure you check local laws.
- If you're having trouble visualizing the boundaries, try using a garden hose or rope to outline the garden spaces or mature tree canopies.
- If you can't afford a full forest at the start, set up a nursery.
- Start composting while the mulch does its thing. You'll then have ready-made fertilizer for your fruit & veg.

THE "LOOK-MOM-NO-LAND" FOOD FOREST

By the way, how would you react if someone told you to create a food forest with no land at all? New York artist Mary Mattingly has done exactly that: her Swale project was built on a floating platform with more than eighty species of trees and plants. As an educational and awareness-raising project, it's been a massive success, visiting different parts of the city shoreline along the Hudson River and sparking off other public food forest projects like the 'foodway' in Concrete Plant Park in the Bronx.

I bet your design problems aren't nearly as extreme as that!

CREATING THE FOREST GARDEN—PLANTING

There is one job that needs to be done first—clean-up. Whether it's taking down a massive half-dead tree that won't stay in the design, taking rubble or old sofas or shopping carts out of the garden, or chopping down a jungle of brambles, get that done first. Taking out the trash may be obvious, but why is the tree coming down? Because if you don't, by the time you want to take it down, you will have a load of younger fruit trees planted and it could fall on them.

In terms of planting, not all the trees have to be planted all at once. You can plant your support species and productive species at the same time, then manage the support species to protect, shelter and boost the productive ones. Or you can just plant the pioneers. But don't plant just the productive plants. Why? Because in the first year, ninety-five percent of the effort goes to creating biomass. You need to get a canopy for shade first. Get the legume and mulch-producing trees in first so that they're fixing nitrogen and protecting and moistening the soil. Get the hedgerow up and running if your garden needs shelter from the wind. You can always chop these trees down later and make them into mulch and hugelkultur beds or timber. This is what we call 'stacking in time.'

That's probably more important in the tropics and in a larger food forest, where you'll need to pick fast growers for those leguminous trees. But wherever you are located, don't miss out the support—the mulchers, the nutrient accumulators—because those are the plants that are looking after the needs of the productive plants for you.

If you've got large trees that you established in the first year or so to protect the fruit trees, you can either fell them for lumber or firewood or pollard them—that is, take them down to a certain height and let them regrow branches from there. That will keep them doing the work of producing mulch or fixing nitrogen, but it will let your fruit trees get more light.

If your fruit trees are your largest trees, though, then they'll be going in first. If you're planting bare-root trees, soak the roots in a bucket of water first and make the hole nice and wet by pouring water in the day before. Then ensure they are well mulched (e.g., sheet mulched with cardboard and then compost) and protected from grass invading their roots.

This may be enough for your first year. Maybe it's enough work, or maybe you want to spread the cost of the plants (which can be considerable). By getting these plants in, you have established the backbone.

If there are perennial weeds you want to get rid of, you could sheet mulch most of the garden, or you could sow annuals like nasturtiums and herbs and salads in the gaps.

You should definitely put some nitrogen-fixers and nutrient accumulators in the ground cover layer to ensure the health of your fruit trees. Just planting comfrey, for instance, and sowing some clover, would really help, even if the rest of the system isn't going to get going till next year.

Another way to start could be to plant your food forest guild by guild and eventually link them up. So you could begin with a walnut guild in the far corner and then add your mulberry barrier guild, then after that, your apple, pear, maybe peach, or plum guilds coming into your garden in roughly a quarter-circle. That means you're only buying one big tree, two or three shrub bushes and some perennials, plus a few packets of seed each year. If money is short, remember you can divide many plants or take cuttings, while others will self-seed, so once your first plant is established, you never need to pay for another.

By the way, you can encourage mycorrhizal activity by buying some inoculant spawn powder and dusting the planting holes before you plant the shrubs.

If you want to plant annuals, you might also use some temporary raised beds. You can build them using chicken wire or old pallet wood, use them for a couple of years and then distribute the soil to reuse for your forest. Or you can plant trees directly in them later on. You might even use a hugelkultur, for instance, if you've cut down an unwanted big tree or a lot of scrubs, and use that for a few years before finally bringing your forest a bit closer.

So there are quite a lot of ways you can build a food forest. All at once, layer by layer, or tree by tree (guild by guild). In fact, when you're planning, you might want to build a load of plastic overlays (get transparent acetate projector sheets and then draw a separate layer of the garden plan on each one) to your basic plan (or overlays in your software), so you can plan how your food forest will grow year by year and how you'll use the rest of the space in the meantime.

Be prepared for things to look horrible for the first couple of years, by the way. During this time, your trees are working hard underground, getting rooted in, driving down for food and water, but they may not be growing much on top. It's often the third year that makes all your work feel worth it: that's when

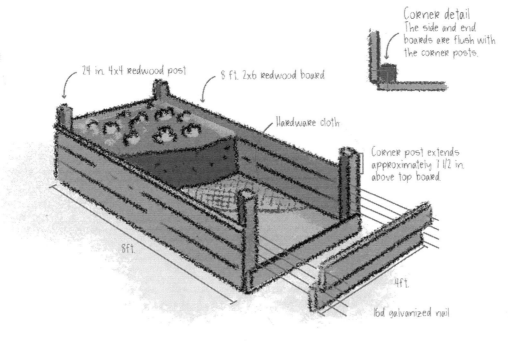

RAISED BED DESIGN

most food foresters find things really get moving and the patch of straw mulch and twig-thin little saplings has suddenly become a wonderful green garden.

Forest gardening is 'no-dig' for most of the time, but it's not 'no-work.' Pioneer Robert Hart talked about occasional 'editing'—unlike with an annual vegetable bed, you only have to edit (prune, harvest, trim); you don't have to rewrite the book every year!

Let's look at how that works with regard to most of the 'gardening' jobs most gardeners do all the time:

Digging? Once you've dug the holes to plant the trees and shrubs, there's no more heavy digging to do, certainly not the yearly double-digging some conventional garden methods use.

Watering? Again, water the plant roots and holes well when you first plant them, but once your forest is established, mulch or ground cover should keep the moisture in the soil (and if you have swales, that will help too).

Weeding? You won't need to do much; there won't be much room for the weeds to invade. Remember, most weed species are pioneer plants that invade bare soil; they will almost always lose if they try to duke it out with established ground cover plants. You might need to just put your gloves on and uproot a few nettles from time to time. Don't worry about annual weeds, like chickweed—it's just the perennials like poison ivy or couch grass that you need to stop before they can get a foothold.

Spraying? As with any other permaculture garden, you've made the forest garden biodiverse, so it will be more resilient than a monoculture to pest and blight. If you made room for helpful slug-eating birds and lizards and added insectary plants to attract beneficial insects, they will do a good job of minimizing pests.

Fertilizing? Well, there's a little work to do here. But if you have plenty of plants like comfrey, it's just a question of chopping them down and turning them into mulch a couple of times a year. You might also bring in some compost and apply it in the winter, when the annuals have died down, giving it time for earthworms and rain to incorporate the nutrients into the soil in time for spring.

Pruning? Occasionally, you will want to prune the larger trees or maybe get rid of one or two altogether to open the forest up a bit to light. Then you can chip the twigs and spread a whole load of that biomass around or start a new hugelkultur. But that will be once or twice a year (depending on your climate), not every weekend. The weekend work? Just walking around, seeing if there's anything seriously wrong that needs putting right if there's anything to be picked and eaten (possibly on the spot) and enjoying the garden.

Compared to my granddad, who was always mixing some chemical or other in his watering can and going out to fertilize the vegetable bed. If you have a food forest, you'll have a much easier life (or should that be a *mulch* easier life?).

EXERCISES

Take a walk on the wild side

Depending on where you live, this may involve walking to the end of your driveway or taking a trip some hours away from home. Go and find a forest, not a plantation, but a natural forest. Take a notebook and a camera or smartphone. Observe the plants and wildlife. What's on the ground—pine needles, wild strawberries, violets, garlic? What different layers can you see? Can you see what happens at the edges? Can you see tracks of animals? What birds and insects can you see?

Turn over a few dead branches. What's underneath? Look at the mosses and lichens. Take pictures of plants you don't recognize and look them up later. What jobs are the different plants doing in the ecosystem? Can you find different micro-habitats?

When you've done the exercise, take a nice long hike, or just sit down somewhere nice and enjoy a moment of Zen.

Take a walk on YouTube

While walking a real forest shows you what works in your particular climate, it can be inspiring to look at food forests around the world. Geoff Lawton has videos showing those Moroccan and Vietnamese food forests I mentioned, as well as his own work at Zaytuna Farm and in Jordan; backyard gardeners have video tours and retrospectives, which are great for seeing how their food forests evolved over time. There's even a video of Robert Hart's forest garden—the first in the UK. You'll get a great feeling for how productive these systems can be and how much sheer joy radiates from their owners.

You might also want to check out Sepp Holzer's 'Farming with Nature' and material by Bill Mollison and David Holmgren.

12

MAINTAINING AND HARVESTING THE PERMACULTURE GARDEN

I mentioned before that permaculture is no-dig, but it's not no-work. However, it's *smart* work. Permaculture gardeners only do the work that nature isn't already doing for them. It's like managing a factory instead of actually making the product.

VISITING

One of the things I most love about a permaculture garden is that I can actually take two or three weeks' vacation and not have to worry about it. But it's important, the rest of the time, that you do have a little wander round every day.

If it's on your balcony, that's not difficult. If it's a bike ride away, it can be trickier to fit in the time. But you don't have to take long. You are just wandering around, enjoying the garden, seeing how it looks, what stage of development plants have reached. If you see a plant that shouldn't be there, you can uproot it. If the plants need water, you can deal with it. If the mulch is looking a bit thin, add a bit more. If the tomatoes are ready, think about spaghetti for dinner and pick a few together with a bit of basil! Sitting in the garden can provide a different perspective to walking around and you may spot things you otherwise wouldn't.

The younger your permaculture garden is, the more important these little tasks are because it can take a while for a garden to get established. For instance, the plume of water downhill of a swale will become bigger over the years; at first, it will only have a small effect on the plants five or six feet away, but later, it will have more impact. So, in the beginning, you may need to do a bit of focused irrigation, but later on, the swale should do most of the work for you.

I can afford to take time off, as I've had my garden a few years now and most of the plants are really well grown. But in the early days, you'll want to be a bit more careful.

WATERING

From my childhood, I remember the pain of coming home from school and having to go straight to water the tomatoes. And we didn't have a hose; I had to carry a heavy watering can all the way down the garden. Every. Single. Day.

Fortunately, permaculture means saying goodbye to the watering can most of the time! There may be the odd occasion when you need to get this out, but it should be the exception to the rule.

If you dig a swale, it's hard work at the time, but it makes life so much easier later. Over time, the swales will accumulate water and hydrate the soil further downhill. If you dig rain gardens or pools, again, you make your life easier in the future. If you mulch deeply, the mulch will keep moisture in the soil instead of letting it evaporate; the same is true where you have ground-covering plants instead of leaving bare soil.

If you have made a good permaculture design, you have also chosen plants that should thrive in your climate and placed all of them where they get exactly the best microclimate for their individual needs. So they ought to thrive—you haven't put them in places where they don't get enough water, sun, or nutrients.

Even so, in very dry climates or in a heatwave, you still may need to water some plants, but not to anything like the same extent that you would in a conventional garden. Sometimes, you may just need to water the youngest trees and shrubs, or most recently, planted plants, which are still getting established; the same applies to seeds while they are germinating. Using a seed blanket, or planting the seeds between two layers of paper or wood pulp material, can keep the seeds from drying out before they're able to sprout (it also stops rain washing them away), although some seeds need light to germinate, so must be surface-sown.

When you do need to water, do it first thing in the morning, so it's still cool. The water can run down to the plant's roots over the next couple of hours and there's not going to be as much evaporation as later in the day when things have warmed up more. Watering earlier also means the water is available to the plants throughout the day.

If, for whatever reason—like a work trip—you have to water in the afternoon, make sure you do so early enough for any water on the plant to have dried before nightfall. Wet leaves can encourage mildew and other fungal problems.

Remember that plants take in water through their roots; that's why tilth and organic matter in the soil are so important. That may sound dumb, but I see a lot of sprinklers in gardens set up to water the plant's leaves, not its roots! That is ridiculously wasteful. Make sure the water gets where it's needed by watering around the roots of a plant directly, rather than watering the leaves as well.

It's better to water once really thoroughly rather than to give plants a shallow watering more often. Frequent, light watering encourages shallow roots and that makes plants more at risk of drought; if they're encouraged to sink their roots deeper, they will be more resilient. If plants are already looking dried out, give them a whole load of water right away and make sure it soaks in. It's the only way to save them.

The one exception to thorough watering is plants grown in containers. They will need more frequent watering, as they don't have as much soil to draw water from as a plant growing directly in the earth. Give them about a tenth of the volume of the container and pour it on very slowly so that as much as possible is absorbed. Don't pour it quickly, or it will just run out of the bottom without being properly absorbed by the soil or compost.

For smaller pots, you can easily gauge how much water the plant has by weighing the pot in your hand. If it feels heavy, the soil has absorbed a good amount of water; if it feels light, it needs watering. Smaller pots need more frequent watering than larger ones, as they have a greater surface-area-to-volume ratio—more opportunity for the moisture to evaporate through the sides if they're porous.

Different plants need different amounts of water. Annual plants, particularly young seedlings, need more water because they only have shallow root systems. Established perennials need less since their roots are more established and they can look further and deeper for water, but trees planted in the last couple of years will need more watering. In fact, it can take a tree up to five years to become fully independent, so giving your younger trees a thorough watering when the weather is hot can save them from getting drought-stressed.

Onions don't need water. Leafy crops need a lot. Tomatoes and potatoes need more watering once their flowers have appeared and they're putting more effort into producing the parts we're eventually going to eat. In temperate climates, most plants grow fast in spring and summer, when they need about an inch of rain a week.

But plants from other climates may need more or less. Desert plants like succulents or cacti want to stay dry. They might need to be watered once every few months, then let them dry out completely. Ferns, on the other hand, as well as tropical plants from a rainforest or cloud forest ecosystem, need much more frequent watering.

Don't let roots get waterlogged. That's particularly important for containers. Some gardeners prefer to let a container stand in water for half an hour, sucking the water up, then take it out and let it drip dry before putting it back in the saucer.

A plant can look as if it's wilting and needs water. But it's possible that it's wilting because its roots are not getting enough oxygen if it has been over-watered. Some plants wilt during the day but revive by nightfall.

Wilt can also be caused by disease. So don't water as a knee-jerk reaction to wilted foliage—make sure you know the real reason before you water.

Self-watering systems can help you keep plants steadily hydrated. Putting a cotton wick in a bottle of water and running it into the plant pot is a cheap way to do this. Seep hoses and irrigation systems also have their uses. However, as a permaculture gardener, you're looking to close the cycle, so you should be using greywater, rainwater, or even pond water. Avoid tap water, as chlorinated water can destroy beneficial microbes in the soil and that can lead to poor soil and even to pest attack.

If you water directly, use a 'rose' as the shower head attachment that goes on the watering can. It spreads the flow, so the water doesn't erode the soil or splash back on to the leaves.

Watering should only be done when it's needed. If you put your finger in the soil, is the soil dry for a whole two inches down? If so, it needs water. But it doesn't need water just because today is Saturday.

PEST MANAGEMENT

If you have properly designed your permaculture garden, you will have built-in a good deal of pest-repellent plants and attracted pest predators. Creeping thyme, for instance, does both jobs at the same time. It's a hundred percent natural; it includes no pesticides that can kill beneficial insects and no neonicotinoids that can kill bees. Other great pest-killer plants include the neem tree, oregano, parsley, lemongrass, basil, mints, pennyroyal, geraniums, catnip, rue, feverfew, most of the alliums (like garlic and chives), borage and asparagus (for tomato pests), anise (against snails), dill, coriander and fennel. The stronger scented a herb is, the less the pests like it! As mentioned in earlier chapters, it is thought that marigolds (in particular Mexican and French) deter nematodes and whiteflies, but their ability to deter other pests has no scientific proof as yet.

Flowers like calendula, daisies, borage and many others will attract beneficial insects. And when you're interplanting a mixture of crops, pest repellents and insect attractors, fewer pests are going to find the crop plants because there's not a single nice big bed of cabbage (or whatever) for them to find.

Always choose the most pest-resistant and disease-resistant variety of a plant that you can. Often, though not always, these will be closer to the wild variety or will be native varieties. Note, however, that disease resistance can often be a trade-off with other things such as enhancing flavor, so a balance needs to be struck between competing wants. Tromboncino squash seems to be the most resistant of the squashes; for instance, with butternut, acorn squash and Improved Green Hubbard are more resistant to squash bugs. Butternut and Green Striped Cushaw are resistant to both squash bugs and squash vine borers. Good varieties of sweetcorn tend to be those with tight husks, like Country Gentlemen, Seneca, Silvergent, Staygold and Golden Security, which are more resistant to corn earworm.

Plant them in the right conditions; a plant that's not in the right position is a plant that is likely to be weaker than it should be and thus more vulnerable to pests. If a crop needs full sun, give it full sun; if it needs a moist environment and shade, find the right niche for it tucked away under a tree or bush.

As well as bugs, there are various kinds of blight and mildew. If a plant gets blight, destroy it. Don't put it in the compost, burn it, or put it in the trash. You may also decide to destroy a plant with mildew in order to stop the infection from spreading to other plants. Note, however, that a small amount of neem oil spray can clear up minor infections pretty quickly, and there are certain plants such as courgettes/zucchini that can get mildew towards the end of the growing season with no noticeable ill-effect.

Rotate annual crops as much as you can. For instance, don't grow potatoes, tomatoes and other related nightshades (Solanaceae) following on from one another—grow salads or beans instead. In the case of blight or other diseases, don't plant the same or related plants in the same place for three years. Other non-nightshade plants can also suffer from blight, such as raspberries.

Keep records—what pests, how many plants were affected, what you did, the outcome, any other circumstances that may be relevant. Identify the pests. Try to work out why the plants were infected or attacked and think of something you could do next time to prevent it from happening again.

You may find that as the soil improves and the microbial life in the soil gets more active, some of the pests disappear or become much more sporadic. That's probably just because your plants are healthier. You know that colleague of yours who is always getting whatever sneezes, flu, or whatever are running around the office? What's the betting that if he paid a bit more attention to his general health, his system would be in better shape to fight off the flu? Likewise, better general plant health tends to bring greater resilience to pest and disease attacks.

Some organic gardeners use diluted essential oils as a spray to protect plants. Clove, basil, mint, citronella, lavender, geranium and eucalyptus are common. But I think planting basil and mint, or lemongrass, is a lot cheaper and just as effective.

HARVESTING

Most packets of seeds show a grid on the back with sowing and harvesting months colored in. You'll notice that most of them show two or three months, sometimes even more. That's quite sensible since every small change in climate can shift your harvesting window. Even a microclimate can make a huge difference; if you grow squashes in a sun-trap formed by a low hedge or figs against a white, south-facing wall, they'll ripen earlier than the same veg or fruit grown in a different place in the same garden.

This is another time that good record-keeping really helps. You may find that harvest happens about the same time every year, or you may find it's always a certain number of weeks after sowing, which is handy to know if you were prevented from sowing at the usual time because of a late frost. If you harvested your first tomatoes on July second, but they weren't really ripe, make a note—leave it a bit later next year, maybe. Note the yields; you can be really geeky and weigh everything, or just note, "this variety of tomato was a disaster," "this squash is a brilliant huge crop."

So in future years, you can look at your records for the right time to start harvesting. But as so often is the case, the best thing is to observe—you will see your vegetables and fruit ripening, changing color, changing size and shape. You can tell melons are ripe by smelling them or just tapping them—you'll hear the difference! You may have to open up a corn kernel to see if the cob is ready—the kernel needs to be plump and wet and the silk completely dry. Or you might, with berries or cherry tomatoes, just pop a couple in your mouth to see if they're ready.

Some crops are choosier than others. You've got just three days to pick your corn before it starts to rot. Onions are more tolerant and root vegetables more tolerant still—some people just dig up their potatoes when they need them; the same goes for carrots and sunchokes. However, if you let vegetables like carrots and parsnip grow too big, they can get a bit tough and lose some of their taste.

Other veg you just need to judge by size, like zucchini, eggplant and squash (yes, you do need to know how big the variety you planted is meant to be, so don't throw the seed packet away!).

Usually, it's best to harvest in the morning before it gets hot. Try not to harvest blight-prone crops like tomatoes and squashes when it's wet. You also need to commit yourself to visiting the garden every day during the harvest if you want to make the best of it; fruit and veg can grow and ripen amazingly fast. And take a good sharp knife or secateurs/small garden shears and a small garden fork so that you don't damage the crop or the plant. Pumpkins, for instance, will rot quickly if you don't harvest them with about three inches of the stalk left, so you need to cut them off the plant, not pull.

Remember that for some plants, you're going to be harvesting them for a while. For instance, a cherry tomato plant can continue to deliver tomatoes for a couple of months. Regular picking will encourage it to keep going. With greens like sorrel and chard and some herbs, like basil, regular harvesting should pinch out the flowering tops and this will keep the plant green for longer before it flowers. Lettuces also 'bolt' (run to seed), so be sure to pick them before this happens, unless you want to leave a few from which to save seed. For some late-ripening fruit and vegetables, like pumpkins, remember to pick what's left before the first frost. And take a good basket or box so that you can treat your fruit and veg gently. Many will bruise or squash if they are roughly handled.

A gardener harvesting in her garden

Vegetables don't just stop ripening when you pick them. If they are ripe, you'll need to chill them, or in the case of some root vegetables, keep them in a root cellar in sand to stop them shrivelling up. If they aren't ripe, like green tomatoes, you'll want to store them at room temperature.

STORING AND PRESERVING

Permaculture is a no-waste philosophy. So don't pick fruit or vegetables you don't know how to use.

It's always nicest to eat your food really fresh. But some fruit and vegetables keep longer than others. Leaves like lettuce and spinach will not keep very long, even in the fridge; cucumber will last four or five days. Strawberries and other soft berries (and figs) will not last more than a day or two. Sweetcorn won't even last that long; you really need to cook it the day you harvest. On the other hand, you can keep winter squash in a cool, dry place for months. Sweet potatoes also store well. Maincrop potatoes store better if the foliage is cut off a week before they are dug to allow skins to thicken. Onion and garlic store better if left to dry out, preferably on a drying rack in the sun for a day or two before hanging up.

Apples and pears can be wrapped in paper and stored in a single layer in a fruit box or on a shelf. Check from time to time that none of them have gone bad, as that could spread to others. Root vegetables like beetroot and carrots store well if covered by a little sand; this stops the skins from getting tough. Potatoes ought to be stored in the dark, in a paper bag or hessian sack.

Suppose you've still got more than you can use right now; you have quite a few options.

Some vegetables need to be 'cured' for winter storage. Harvest onions only once the stem has turned brown and flopped over. Then they need to be cured, either by just laying them down in the garden for a couple of weeks, if it's dry, or lay them out in a single layer on racks in a warm shed, loft, or

When to harvest various fruit and vegetables	
Fruit	When to harvest
Apples	Ready to pick when a gentle twist will part the stem from the branch without tugging
Asparagus	March onwards in US and Europe; harvest as soon as the spears are about eight inches long, before they begin to break into leaf. Cut just below the soil
Beans	Harvest when you can see the pods are well filled out, but haven't started to yellow
Beets	About two months after planting; baby beets can be harvested earlier (and don't forget the leaves can also be eaten in salad or stir-fry)
Blackberries	The clue is in the name; when they're black (if they're still red, they're not ready)
Cabbage	Cabbages are ready when they have a tight, firm head. Early varieties may be ready three and a half months after sowing; late varieties take five to five and a half months
Carrots	Carrots can be harvested from bite-size baby upwards. If they're sweet, and crisp, they're ready. Remember to use a garden fork to loosen the soil before you pull them up
Cucumbers	Pick your cucumbers a little on the small side for better flavor
Eggplants	Harvest when the skin is nice and glossy, and don't leave them on the plant too long
Figs	Should be squeezable but not squishy. They will also start to hang downwards when they're ready for picking. Depending on your climate you may get two crops a year, or one
Garlic	Ready three or four months after planting. Once the tops have started to yellow, you can pick the bulbs; they should be stored in the shade to dry. Traditionally, they were often stored in bunches made by plaiting the stalks
Jerusalem artichokes	The foliage will blacken and die in autumn; this is your cue to start lifting the tubers. You will need a garden fork to loosen the soil
Melons	Cantaloupe is ready when it will fall off the vine easily when the stem is pressed. Honeydew is ready when the skin has gone bright yellow. Other melons are ready when they smell right. And watermelons? If tapping the fruit makes a dull sound, they're ready

When to harvest various fruit and vegetables	
Fruit	When to harvest
Peanuts	Lift when the foliage has gone yellow and the pods have filled out (if they're still not ripe, come back in a week or so as they continue to ripen after the leaves die off)
Peas	Usually take about three months from sowing, or three weeks from flowering. Don't let the pods go yellow and dry unless you want seed or dry peas
Peppers	Peppers are fascinating vegetables as you can harvest them green, yellow or full red - they will ripen on the plant and taste completely different depending on when you pick them
Raspberries	When fully red. When they're ripe they will practically fall off the vine in your fingers; if you have to tug, they're not ready
Tomatoes	For cherry tomatoes, the taste test works. For larger tomatoes, pick when there is just a little 'give'. It's better to pick unripe rather than overripe - you can put tomatoes on the windowsill to ripen. You can also use green tomatoes in chutneys and salsas (useful at the end of the season)
Strawberries	When they are fully red; they'll also smell delicious when they're ready to pick
Sweet potato	Dig your sweet potatoes up after the first frost. If you are lucky enough to live in a frost-free region, about three months after first planting should see the first potatoes ready to harvest
Zucchini	Don't let them get too long - four to six inches is enough. They can split if left on the plant too long

porch. Clip the stem about an inch from the bulb—if there's any green left, they need a bit more curing, but if the stem is completely dry, they're ready.

Do check daily to make sure if any onions spoil, you throw them out before they start affecting the others.

Potatoes just need to be left in the ground for a couple of weeks after the foliage dies and then put in a dark, moist environment—somewhere humid like a cellar, but a well-ventilated one. Leave them there a week or so.

Drying is one that I like because it retains a lot of the flavor. Either sun-dry on trays, or if the weather isn't good enough for that, buy an electric dryer. You can also set your oven to its lowest temperature (not more than 250F or 140C) and dry the crop for several hours—it works best with a fan setting if you have one. Sliced apples, halved tomatoes and peppers, quartered pears and other fruit can be dried easily.

Freezing is a good way to preserve your harvest. Most berries freeze very well, though strawberries will not retain their crispness. Some fruit and veg need blanching—putting in boiling water for a few minutes, then quickly rinsing with cold water—before bagging them up for the freezer; beans, apples, broccoli, okra, leafy greens and asparagus should all be blanched. Freeze in the right-sized portion for you and your family to use at a single meal.

Some gardeners I know make big cauldrons of different vegetable soups and huge batches of fruit crumbles and pies that they then freeze. That can be good if you have spare time around the harvest season (and freezer space). Again, just make sure that the portions are the right size for a meal for you and your family or friends, not too big to use up. Eating the same soup every day for a week is not anyone's idea of fun.

I also freeze herbs like coriander and basil using ice trays. I chop the herbs, then freeze them in ice cubes. Or you can make herb butters by blitzing herbs with butter and storing them as a roll (sausage) or in cubes; those freeze excellently and are really easy to use in the kitchen.

Pickling and jam-making are other ways to keep your crop usable for longer. Both will need you to do a bit of learning, but if you enjoy cooking, making your own conserves can be great fun. Use your imagination—my favorite jam now is my homemade fig, orange and cardamom jam made with the very last figs of the year. I spent three years tinkering with the recipe and finally got exactly what I wanted! I'd rather spend time in the garden than the kitchen, to be honest, but that's a damn fine jam, so that's one day a year I really do want to get the saucepan boiling away.

Some fruit also preserves well in alcohol—you'll need vodka, brandy, or bourbon that's pretty strong, plus sugar to taste.

But besides thinking about techniques, let's also think about 'fair shares.' What about sharing your harvest with friends and family? Or donating to a local food bank or charity? Or, if you can't pick all the berries from your edible hedge, what about sharing with the birds? That will keep your birds fed over the winter and they'll be ready to take up their duties of pest control and fertilizer making next spring.

HOMEMADE JAM

FALL CLEANING

In the house, we do spring cleaning. In the garden, we do fall cleaning—at least, in temperate climates. Things may be different if you have a monsoon season.

As with harvesting, one thing you should do is to take notes and make records. Don't bet you will remember what you planted and where; if it's below ground, or if you have two different varieties of a fruit tree, you'll forget where it is and which is which. It's a good idea, too, to sit down and jot down notes on how the season went. What did well? What didn't work? What surprised you? What did you enjoy?

Chop down the dead plants. You can use them as mulch where they are, or you can take them to be composted. Don't bother pulling out the roots; if they stay in the soil, they'll rot down gradually, add to the organic matter in the soil and also help the soil structure. Plants that have had any disease, on the other hand, need to be pulled up entirely and then put in the trash, not the compost pile—or if you are in the country, you can burn them.

Have a walk around the garden and, for once, do a little weeding. If you've got a few too many dandelions or a little too much chickweed, chop the leaves—again, leave the roots to do their work. If you see any really nasty running weeds like couch grass or ivy, pull them up and destroy them (unless you have a really, really hot compost heap that will do the job).

Loosen the soil a little with a digging fork. That will help the fall and winter weather break the soil up. But don't walk on the soil or compact it. You can then add some finished compost, preparing for the spring planting by giving the soil a bit of extra fertility.

You'll then either want to sow a cover crop or mulch the beds. For instance, crimson clover or winter rye work well; they'll prevent the soil from being eroded or waterlogged and keep weeds out over the winter, then in spring, you can use them for mulch or compost. You can mulch with straw; if that doesn't look tidy enough for your liking, scatter wood chips or bark on top; or you can use leaves that have been composted for a year.

If you want to plant new beds in spring, this is a good time to build them up, mulch and compost them and get the soil in good condition. And finally, when all this is done, clean all your garden tools properly and store them indoors—whether that's in your mudroom or a garden shed—away from rain and wet. Alan Titchmarsh always recommends having a bucket of sand with oil mixed in. Dunking your tools in will both clean them and coat them with a protective oil for waterproofing.

EASY SEEDS TO SAVE

Saving your own seed instead of buying new seed every year is an excellent way to cut the cost of your permaculture garden *and* to preserve older or different varieties of vegetables. Lots of organic and permaculture gardeners enjoy swapping their heritage variety seeds and this is a kind of sharing that's fun to do, as you can try new varieties.

Remember, though, that F1 hybrids—often the most heavily marketed vegetable seeds—won't grow true to type. You need to grow open-pollinated fruit and veg to be able to save seeds viably.

Also, be aware that if you're not careful, some plants will cross-pollinate and you'll end up with a mix. If you grow different types of sweetcorn and squash, they may cross-pollinate. If you're really set on saving seed, you can pollinate a small number of the female flowers yourself with a small paintbrush. Take the pollen from the male plants on the same vine and transfer it to the female flowers, then cover them with a paper bag till the fruit has begun to grow (don't use a plastic bag—it will overheat and can also create problems with moisture).

With other vegetables which have more, smaller flowers, you'd take far too much time pollinating each flower yourself. So the best course is to separate different varieties.

Harvest seed from the best and most vigorous plants with the best-tasting fruit or veg. You're naturally helping to keep the best and not the worst of the plant's characteristics.

With tomatoes, pick the ripe tomatoes just as you would to eat them, scoop the seeds out, wash the pulp off them and put them in a jar of water for three days, somewhere warm. That ferments the pulp away and kills off any germs. Then pour the water and seeds out onto a paper towel and let them dry. When they're fully dry, put them in envelopes and don't forget to mark which is which (I learned that the hard way!). You can save the seed from pumpkins and squash in the same way.

With peppers, you want to wait till the peppers begin to shrivel up a bit and the skin wrinkles. They'll be past their best for eating, but the seeds will have ripened properly. Just take the seeds out of the peppers with your fingers and again, dry them on a paper towel.

Other plants will form seed-heads, which you can preserve, like sweetcorn, or pods, like beans and peas. For peas and beans, let the pods ripen on the vine until they're dried out and start to turn brown. Harvest when you have had good weather for a while—you really don't want the pods to be at all damp—then spread them out indoors to dry out more, for at least a fortnight before you open the pods up and save the seeds. In fact, if you want to, you can leave the pods unopened till you're ready to plant them next year.

Lettuce and other salads 'bolt': the center grows tall and creates many tiny, yellow flowers, which eventually go to seed. When they look like mini-dandelions, it's time to pick the seed heads. Unfortunately, lettuce that bolts is bitter, so harvest most of your lettuces before they do and just leave a few for seed.

Chard is another vegetable that's easy to save seed from and re-sow or start off a new area. It will self-sow, but you might want to be a bit more disciplined about where it goes.

Herb seeds will grow once the herb flowers and bolts. Not all herbs will reproduce best by seed—perennials like lavender and rosemary do better with cuttings. Basil, lovage, caraway, fennel, cilantro and dill all produce flower heads that will dry off; cut the whole flower head and shake the seeds out over a piece of paper (it's best to put the flower head in a paper bag before you cut it, to prevent the seeds falling on the ground). Parsley is the same, but because it's a biennial, it will only have flowers after two years, not in the first year.

Some herbs, like chamomile, will self-seed extensively. But there's no harm in taking a few seeds which you can share with other gardeners, or use for 'guerrilla' gardening, or reseed if there's a very hard or wet winter and the seeds in the ground didn't grow.

Most annual and herbaceous perennial flowers also have seeds that can be saved. Daffodils and rhizome plants like iris also produce seed but may take a little longer to get established. Seed heads usually ripen two or three weeks after the flower has bloomed and they'll be either dry and hard or fluffy (like a dandelion). Saving seed is a great way to expand your garden. If you started with a $3 pack of seeds, you'd get one year's flowers. Then you'd need to spend $3 next year and if you'd started another guild that needed the same flower, another $3... and so on. The larger your garden, the more expensive it gets.

With seed saving, you're saving money and you can extend the amount of ground covered by your flowers by five to ten times every year without paying any more than for that first packet. But remember, with a lot of seeds you'll be racing against time, because the birds will be interested too!

SEED STORAGE

You may be used to getting neat little black seeds in their packets. Rubbing most seed heads with your hands will separate the chaff from the seeds. You can put the mix of chaff and seed through a strainer or toss it up in a dish and let the chaff float away (it's lighter than the heavy, dense seeds). But you don't have to do any of that—if you have room, you can save the whole seed head in a paper bag. Vetch has little pods, just like peas, only smaller; lupines are the same.

Most species will keep for a year with no problems. Some will keep longer. A few flower species need to be refrigerated ('cold stratification')—lupine, for instance, as well as some varieties of coneflower, wild geranium, delphinium, and catmint. All seeds need to be kept somewhere cool—as long as it's not freezing. Don't put your seed tin on top of the radiator! Remember to keep track of your seeds by age and always use the oldest ones first.

Seed viability	
Seed	How long will seeds keep?
Onion, parsnip, parsley, spinach	One year
Corn, okra, chives	Two years
Leeks, carrots, rutabagas, peas and beans	Three years
Peppers, pumpkins, squash, melons, basil, chard, turnip	Four years
Brassicas, beets, tomatoes, celery, lettuce, eggplant, radish, cucumber	Five years

Most seeds need to be stored in an airtight container and make sure it's located somewhere that doesn't get damp. A small glass bottle or jar is probably best; a plastic jar or zip-lock bag is next best. Store them in the dark. I use all my medicine bottles for seeds and keep them all in a small, wooden box in the spare room, which we don't heat unless we have guests. My vitamin capsules come with little silica gel desiccant packs, which are helpful for ensuring seeds are kept really dry. I'd recommend you keep them in the house, not in a shed, where mice might try to get at them. Don't forget to label your seeds with the variety and the date you collected them. You can then keep a record of which of your seeds did best next year.

By the way, the oldest seed ever germinated was *Silene stenophylla* recovered from the permafrost in Siberia. It was 32,000 years old and grew at the same time as Russia was inhabited by mammoths and woolly rhinoceros. The next oldest seed successfully germinated was a comparative youngster at just 2,000 years, a date palm seed that had been discovered at Masada, Israel.

Basic maintenance tasks	
Watering	Swales, greywater, mulching, ground cover, etc., should mean there is much less to do in a permaculture garden.
Weeding	Ground cover plants should keep unwanted plants out (plus, sheet mulching will have got rid of the worst nasties), so there should be relatively little to do.
Pest control	Again, little to do if your companion planting and predator/parasite insects have done it all for you. React fast if you see blight or other diseases; pull the plant out and burn it or put it in the trash.

Basic maintenance tasks	
Harvesting	Easy when you have remembered to keep regular pickables (salad, tomatoes) in zone 1. Also easy to get free help for major fruit harvests!
Preserving	For harvest gluts, you need to take time drying, preserving, salting, pickling, freezing… or sharing!
Tidying	Add mulch when it's getting thin. Remove dead annuals and compost them. Prune trees if necessary.
Fall tidying	Loosen the soil a little. Weed. Chop down dead plants. Mulch beds or sow green manure. Save seeds. Make notes on the year—what worked, what didn't, best varieties to grow, what you planted and where.

EXERCISE

The exercise for this chapter is called "Just do it."

Go out and garden! Check which maintenance tasks are appropriate for the time of year and get going with some of those things!

13

MAKING MONEY FROM YOUR GARDEN

Making money may not be relevant to your particular vision for your garden, but it is certainly an option for permaculture gardeners. Organic produce from local producers sells well to customers who are concerned about artificial pesticides and fertilizers, so there is no reason you shouldn't make an income from your garden or even make it your main job.

PERENNIAL CROPS

Most perennials make fewer demands on the soil than annuals and are low-maintenance plants. However, it may take a few years till your perennials are cropping reliably and yielding enough to be worth selling, so this is not a short-term option. You will usually only get one harvest a year, so you need to stagger your crops if you want to be able to sell produce all year round.

For instance, rhubarb and asparagus are ready early in the year. As the year moves on, various different fruits become available—first redcurrants, gooseberries, then cherries, then apples and pears and nuts. If you have room for several varieties of the same fruit tree, staggering them between early, mid-season and late varieties makes good sense.

Be careful with fruit like cherries or apricots where the season is short and the fruit has to be sold and eaten quickly. They're trickier to handle than apples. You'll need to be able to pick a lot of fruit at one time, treating it gently so it doesn't bruise and you'll need to have your customers ready to give shelf space to your crop and create a bit of excitement around it. You'll also need to be able to deliver it as fresh as possible so that it's not starting to rot before it gets to the shop. With a couple of trees, the harvest can go like clockwork. But be realistic about how much fruit you can handle, and make sure you plan your working diary around the harvest so that you have enough hands for the work!

Perennial herbs are also great sellers. An established rosemary bush, lovage plant, or carpet of thyme is low maintenance; people who want to make their cooking 'zing' are always interested in fresh herbs. You may be able to sell to local restaurants, which need to buy in quantity.

Asparagus is particularly useful as an income-generator because not only is it a relatively early crop, it's also a gourmet vegetable that sells at a premium. It can also be blanched by earthing up to exclude light in order to produce white asparagus, which sells at a premium.

In hotter climates, you'll have the opportunity to harvest crops such as kiwi fruit and other vines.

If you're selling unprocessed, whole produce, you're generally going to be exempt from health department inspections and the need for permits. But check with your local authority. Remember that even if you don't need permits unless this is a really small-scale venture (say, selling at a couple of events a year or putting an 'honesty box' in the front yard), then you're going to need to register as a business.

ANNUALS WITH A LONG SEASON

Annuals that are easy to grow and get to market include garlic, onions, potatoes and sweet potatoes. If you have the right climate, you can get an early crop of garlic, replant and sell your second crop later in the year. Selling it tied in bunches gives a great home-grown vibe. Garlic will store well, too, so unlike fruit, you don't have to sell it fast.

Salad leaves, on the other hand, are hard to get to market before they perish. If you are working full time on your permaculture farm, then fresh salad mixes could be a premium seller and enable you to sell packages to restaurants and delis, but if you do one farmer's market a week, you may be better off concentrating on longer-lived produce.

MAKE MONEY FROM SEEDLINGS

Don't forget that as well as selling annuals for eating, there is a good market for seedlings. Many gardeners don't want the bother of growing their annual crops from seed, so they want to buy tomato, pepper, squash and zucchini seedlings ready to put in the ground. If you have saved seed and use your own compost and paper cups or recycled newspaper 'pots,' your cost is practically zero. Basil is an annual herb that a lot of people like to buy ready to put in a pot—it's such a fantastic herb when it's fresh and so boring when it's dried!

You can also propagate perennials from cuttings. For instance, vines, raspberries, strawberries grown from the runners (the little plants at the end of a long stalk) and rosemary can all be grown this way. Or you can divide clumps of perennial herbs like mint and lemon balm.

You'll probably be charging $3-5 a pot depending on the stage of development of your seedlings and the rarity of the varieties you offer. It may not make huge money, but if handled properly, it will help your garden budget, particularly if you grow your plants from seed anyway. You're just growing a few more.

You can sell simply by telling all your friends on Facebook or using WhatsApp to let them make their orders ahead of time and run a pickup operation when the plants are ready. Or you can take your plants to a farmers' market or craft fair. However, there may be a bit of paperwork to do.

And if you have some leftovers, why not donate them to a community garden, or your local church, or some other worthwhile venture?

VALUE VS. SPACE

Remember that the best crops for making money are those with the highest market price compared to the space it takes to grow them. Asparagus, cherry tomatoes and scallions do particularly well, as do some exotics. On the other hand, pumpkins take a lot of space and sell for relatively little.

Spinach, lettuce and greens like pak choi, Chinese leaf and mizuna can be very profitable. But the problem with these plants is getting them to market before they spoil. You're going to need to be picking all the time. Broccoli (particularly sprouting broccoli) and kale might be a better choice if you don't want your garden to be a full-time job.

CUT FLOWERS

If you have attractive flowers in your garden, selling cut flowers can generate good income. Farmers' markets, food markets, restaurants and florists are all potential customers. You'll need plenty of sunshine and you'll also need to be able to plan for flowers all spring and summer long. A mix of annuals and perennials may work best.

There is also the potential to sell edible flowers to be used as cake decorations and in salads (such as violas and nasturtiums), but you will need to ensure you have a well-organized supply chain because, like salad leaves, these are perishable and won't last very long once cut.

Good perennials for bouquets include aster, coneflower, daisies, delphiniums, peonies and all kinds of bulbs like daffodils, maybe tulips and iris. Once you've picked the asparagus for eating, you may have stalks that grow high and feathery—they are much loved for bringing some greenery to a bouquet (as are bolted fennel and alliums). For annuals, snapdragon does well, in different colors; it'll self-seed if you let it, too. Sunflowers are also very popular, as are salvias.

Make sure you have blooms ready for events such as Easter, Mother's Day and so on. Valentine's Day, coming so early in the year might be difficult. Another way to sell is to offer a subscription service with fresh flowers every week over the spring and summer, appealing, for instance, to offices or to bed and breakfasts.

MAPLE SYRUP

If you have a large food forest, it's worth making room for some maple trees. Maple syrup is very labor-intensive, but it's a short season of hard work, rather than all year round (and maple is also one of the most beautiful and useful timbers, should you want to cull a few of your trees).

However, you're going to need to comply with regulations. Most US states have regulations on exactly how it should be processed, will want the sugar shack to be built and cleaned in a particular way and will require you to follow federal labeling rules. In some states, production under a certain level isn't regulated, but it might not be worth your while.

You'll also need to invest not just in your trees but in equipment to collect the sap and to boil and bottle it.

MAKING MAPLE SYRUP

TIPS FOR SELLING YOUR PRODUCE

There are various ways you can sell your produce, depending partly on how much effort you want to make (part-time budget-helper to full-time job) and partly on the size of your enterprise—it doesn't make sense to try to make a suburban garden support you full time. You may decide to adopt a number of different methods, diversifying your routes to market as well as your crops.

One of the easiest ways to start is selling at farmers' markets and other similar events. You don't even have to take a stall yourself if you know someone who is prepared to take your produce, for instance, a fruit and vegetable seller who would like your cut flowers to add to the stall's product line.

But let's take a step back. Think who you want as your customer. You might want to sell to individuals, or you might want to supply restaurants. Restaurants will need regular, fresh supplies and will be demanding on quality, but you can pick to order and you won't have any waste. Do you want to sell just in your own neighborhood? One way that works really well is a box scheme; you deliver a box of vegetables, fruit and nuts every week, at a fixed price, with whatever's in season. You do need a good variety to do this, though—if your customers end up with just tomatoes and lettuce for four or five weeks, they're going to get fed up.

You could also put together 'specials' like an exotic vegetable box with recipes and cooking instructions, a salsa box, or a fresh herb bouquet every week. That could also sell well at a market.

You also may be quite proud of your garden. If you have a large food forest out in the country, you might advertise pick-your-own days for families. It's a great activity for children and it's also a chance for you to educate people about permaculture and the environment (remember 'each one teach one').

Even if you're working on a very small scale, build up your reputation so that when someone new moves into the neighborhood, they'll hear about you. Build your brand—it might be very personal, based on your own personality and beliefs, or it might be based around the land you farm or the type of produce you grow. Be a good neighbor—you might even decide not to make income for yourself but to do so for local charitable events, or let the local football team auction off a year's produce for a family of four (as long as you're sure you can deliver).

Marketing can be as simple as handing out free samples, or you can hold workshops to explain permaculture principles. You might give tours of the garden—either in person or online—telling the story of how you started, what you're growing, where you see your garden taking you in five or ten years' time. You might blog about your garden, Instagram pictures, run a Q&A. One city gardener where I live had an open day and put up a whiteboard that said: "What produce can't you get around here that you miss?" He ended up growing a few plants he hadn't thought about before, like yams, sweet potatoes, watercress and mustard greens. Unfortunately, he couldn't do anything for the lone Cajun who wrote "shrimp."

REGULATIONS

You will need to check up on what the regulations have to say. For instance, if you want to have an open day, does that constitute a 'nuisance' to your neighbors? One reason I haven't talked much about options such as making jams and preserves for sale is that the moment you start processing food, most countries start wanting to inspect your kitchen and ensure you do things a particular way. It may eventually become an option you'll want to explore, but if you do, it's a whole new adventure and you probably ought to treat it as a separate business.

You'll also need to set up a business structure, whether that's as a sole trader or if you decide to incorporate. You probably want some form of public liability insurance; if someone comes on an open day, slips on mud and breaks a leg, you'll be covered. You may well be capable of keeping your own accounts, but it's worth speaking to a local accountant or your chamber of commerce about structuring your business properly. That's particularly the case if you're buying a site for a fully commercial permaculture farming business, as there may be advantages to buying through a company rather than personally.

Don't forget to keep your accounts and don't forget to check what's profitable and what isn't. As so often, keeping adequate records will stop you from making mistakes.

DOES IT MATTER?

For some people, permaculture isn't about making money. It's about doing their thing to help the planet, about establishing a garden that's a haven for their soul, or about educating others.

One gardener I know donates extra produce to local charities and lets it be known that if anyone in the neighborhood is short of good, fresh food, they only have to ask. Some people have too much pride to take things for free, so she also accepts a couple of hours' work on the garden as 'payment.' Other permaculture practitioners belong to local exchange networks which let them swap their produce for someone else's work—a free IT checkup, a bit of home maintenance, or some help with homeschooling their kids.

Of course, your decisions might change over time. If you're buying a big property and planting a food forest, you might have decided you'll take eight years growing the forest and then you can give up the day job and work full time on your farm. In that case, you should probably do a little trading in your early years to make sure you have the feel of the market and know what sells and how much work is involved.

But no one is going to judge you for not making money out of your garden—you decide what's appropriate.

EXERCISES

A trip to the supermarket

Take a note of everything you grow, or plan to grow in your permaculture. Do you know how much it sells for? Off you go to the local supermarket and find out! Or try the online ordering system to find out prices for fresh vegetables and fruit. You might also want to look at Whole Foods Market, which sells organic products. Did you have any surprises?

If you know roughly how much you produce, you can now run a quick spreadsheet that tells you what you'd get if you sold everything you produce with no wastage and without eating any yourself. This is not a business plan, but it *is* a very interesting number because it may shock you just how little your garden is worth—or, alternatively, how much!

Estimating your annual food bill

If you're a well-organized person, you may know how much you spend in a year on food. Or you may know how much your main monthly shop is costing you.

In any case, it's definitely worth working this out. Then think about what you can cut out by growing it yourself. Beef, maybe not. Wheat flour, making your own spaghetti, maybe not. But salads, fruit, nuts, peas and beans? You could be saving yourself $600 a year just on an average 600 square foot garden—even just growing your own favorite herbs on a balcony could save you $100 or more every year.

Even better, your food will cost you zero air miles and zero carbon and isn't pumped full of fertilizers and pesticides!

CONCLUSION

This book has been a whistle-stop tour of permaculture. I've aimed to give you the basics that will get you started, wherever you live (maybe not Antarctica, but pretty much anywhere else).

These central concepts of permaculture are founded on the insight that natural systems are highly productive and, at the same time, self-sustaining. The Amazon doesn't depend on someone flying over it with a load of pesticides every year or adding fertilizer every couple of months. By copying the natural system, we get a self-sustaining system that doesn't need a lot of human input; and we can tweak it a bit to provide us with the plants we want to eat or use in other ways.

Nature has its own patterns, which we try to copy. Most importantly, we start to look at plants in terms of their functions and their relationships with other plants, not just with us. We don't grow daffodils in the orchard just because they're pretty, but to stop weeds and grass invading the space that apple trees need for their roots. Sweet potatoes can make a good ground cover as well as giving us lip-smacking potato wedges. There are plants which fix nitrogen in the soil, plants that bring nutrients to the topsoil, plants whose leaves hide the earth and keep it moist and cool for other plants' roots—it's a cooperative community and a dynamic one.

It's a big change in mindset from the way gardening catalogs and lawnmower salespeople want you to think! That's why we had to spend a couple of chapters explaining the philosophy and the theory of permaculture; it doesn't come naturally to most people unless they've studied ecology or worked as a ranger in a rain forest (which was the background of one of the founders of modern permaculture, Bill Mollison).

In a good permaculture design, all the pieces of the jigsaw fit together; the analysis of your site in terms of the energies coming into it, like wind, sunlight and water and how that affects what you should plant where; analyzing your soil—which is as alive as the rest of the garden—and getting it healthy and productive; and building relationships between plants, creating guilds and eventually, full-scale food forests.

You've learned how to design a permaculture, even if all you have is a balcony or a tiny yard. I hope you've learned to keep your expectations realistic. A permaculture isn't an 'order it, and it shall be done' garden; you might need to plan your installation over several years, piece by piece, depending on the size of your plot and of your budget. You'll definitely need patience; natural forests take fifty years to grow and even though we can accelerate things quite a lot, it will still take five years or more for a food forest to become mature.

You've learned about maintenance, too, and I hope that one of the things you've learned is that a permaculture garden is far lower maintenance than a conventional garden. There will still be some 'editing' to do, but lawn-mowing, weeding and watering will no longer consume every weekend.

We spoke a little about making money from permaculture. While many people are satisfied to adopt the principles in their own garden and have an abundance of good food for themselves and their family, some people want to fund their ecological lifestyle and it's a hundred percent possible to do so.

Permaculture is bigger than just gardening. It's about caring for the planet and caring for others; it's a different way of doing things from 'big agriculture' monocultures, which are resource-hungry and deplete the soil. Permaculture saves carbon miles, at the same time as saving you money and helping in its small way to restore the planet.

So now you have all the tools to set up your own permaculture. Depending on your own particular climate, you may need to do a little more work to identify the species that are going to do best there, but you know how to do this and how to build those plants into companions, guilds and food forests. I hope the exercises at the end of each chapter also helped you learn factual, useful things and sparked your imagination and creativity too. You have the tools, so get out there and use them!

And remember 'each one, teach one.' Teach your neighbors about permaculture. Teach your friends, teach your kids, teach that poor guy you live opposite who spends all his weekends sitting on a lawnmower. And if you enjoyed the book, leave a review on Amazon to reach out to others who're just thinking about taking their first steps in permaculture.

Good luck and remember to have fun along the way!

Harvesting Your Permaculture Wisdom:
Share Your Thoughts on "Creating Your Permaculture Heaven"

Now that you've become a Permaculture Pro, equipped with all the tools to turn your garden into a vibrant paradise, it's time to pay it forward. By sharing your honest opinion of "Creating Your Permaculture Heaven" on Amazon, you're not just reviewing a book – you're sowing the seeds of knowledge for other eager Permaculturists seeking guidance in creating their own green havens.

Your Opinion Matters:

Your review is a beacon for others, guiding them to the wealth of information and passion for permaculture found within the pages of this book. By taking a moment to leave your thoughts, you're contributing to a community of green thumbs who help each other flourish.

How You Can Help:

1. Scan the QR code below to leave your review on Amazon.
2. Share your experience and insights – be the guiding light for someone else's permaculture journey.

Why Your Review Matters:

- Your words help others discover the joy of permaculture.
- You pass on the torch of passion for sustainable gardening to fellow enthusiasts.
- Together, we keep the spirit of permaculture alive by sharing our knowledge.

Thank You for Your Contribution:

Your dedication to nurturing your own permaculture journey and helping others deserves applause. By leaving a review, you're playing a crucial role in fostering a community of permaculture enthusiasts who support each other.

Scan the QR code below to leave your review (note this is the link for Amazon US, if you live in a different country, simply change the .com to the one for your country):

Thank you for being part of this green revolution and for making "Creating Your Permaculture Heaven" a stepping stone for others to embark on their permaculture-filled adventures.

Gratefully,

Nydia Needham

APPENDIX

USDA ZONES

The system of Plant Hardiness Zones was developed by the United States Department of Agriculture and is a geographically defined area in which a specific category of plant life is most likely to thrive. They are estimations only and do not take into account any microclimates an area may have. These zones may also change as a result of climate change.

Zone	Minimum Annual Temperature (°F)
2	-50 to -40
3	-40 to -30
4	-30 to -20
5	-20 to -10
6	-10 to 0
7	0 to 10
8	10 to 20
9	20 to 30
10	30 to 40
11	Over 40

Table 1—useful plants

Type of plant
CL—climber
D—deciduous
E—evergreen
SH—shrub
T—tree

Edible part/use
SE—seed
SA—sap
YS—young shoots
IB—inner bark
FR—fruit
LE—leaves
BU—buds
YO—young stems
YL—young leaves
R—root
S—shoots
O—oil
BA—bark
CS—cinnamon substitute
PO—pods
FL—flowers
SP—seed pod
MA—manna
F—flavouring
CA—catkins
YFB—young flower buds

Soil type
LS—light soil
LL—light loam
MD—moist, well-drained
DM—deep moist soil
LO—loamy soil
DS—deep soil
HDS—heavy damp soil
SS—self sterile
MS—moist soil
DL—deep loam
LF—lime free
M—moist
WD—well drained
ML—moist loam
SL—sandy loam
LiS—light sandy
DSL—deep sandy loam
P—peaty
CH—chalk
HR—humus rich
L—lime
SL—sandy loam
D—dry
DE—deep
R—rich
CDS—cold damp soil
C—cool
NTF—not too fertile
F—fertile
DCH—dislikes chalk
A—acid
DP—dislikes poor soil
DPE—dislikes peat
CL—clay
S—sandy

Sun
○—prefers sun
⊕—prefers partial shade
●—prefers shade

Wildlife value
Hab—Habitat
P—food for chickens
F—animal feed
Hum—attracts hummingbirds
B—attracts beneficial insects

Common Name	Botanical Name	Type	Edible Part or Use	Insectary	Wildlife Value	Light	Soil Type	USDA Zone
Alligator juniper	Juniperus deppeana	E.T	FR		Hab, F, P	○	L	7-9
Almond	Prunus dulcis	D.T	SE.O		Hab, F	○	WD.L	6-9
Alpine bearberry	Arctostaphylos alpina	D.SH	FR	B	F, P, Hum	○⊕	LF	2-6
American beautyberry	Callicarpa americana	D.SH	FR	B	F	○	WD.SL.CH	7-10
American hornbeam	Carpinus caroliniana	D.T	SE		Hab, F	○⊕	Most	3-9
American persimmon	Diospyros virginiana	D.T	FR		Hab, F	○	WD.LO	4-8
Angelica tree	Aralia chinensis	D.SH	YS	B		●	DL	4-8
Aniseed tree	Illicium floridanum	E.SH	F	B		●	WD.SL	6-10
Apple tree	Malus domestica	D.T	FR	B	Hab, F	○	WD.SL.CL	3-8
Apricot	Prunus armeniaca	D.T	FR.O	B	Hab, F	○	L	5-7
Aristotelia serrata	Aristotelia serrata	D.T	FR	B	F	○⊕	M	7-10
Azarole	Crataegus azarolus	D.T	FR	B	Hab, F, P	○	Most	5-9
Bay tree	Laurus nobilis	E.T	LE			○⊕	MS	8-10
Bead tree	Melia azedarach	D.T	LE	B	F, Hum	○⊕	Most	7-12
Bearberry	Arctostaphylos uva-ursi	E.SH	FR	B	F, P, Hum	○	LF	4-8
Beech	Fagus sylvatica	D.T	LE.SE.O	B	Hab, F	○⊕	Most	4-7
Bilberry	Vaccinium myrtillus	D.SH	FR	B	F, P	○	A.P	3-7

Common Name	Botanical Name	Type	Edible Part or Use	Insectary	Wildlife Value	Light	Soil Type	USDA Zone
Bird cherry	Prunus avium	D.T	FR	B	Hab, F	○	WD	3-7
Bitter orange	Poncirus trifoliata	D.T	FR.YL	B	Hab	○	Most	6-9
Black hawthorn	Crataegus douglasii	D.T	FR	B	Hab, F, P	○	Most	4-8
Black huckleberry	Gaylussacia baccata	D.SH	FR	B	Hab, F	○⊕	WD.HR.LF	5-9
Black mulberry	Morus nigra	D.T	FR		Hab, F, P	○	WD	5-9
Blackberry	Rubus fruticosus	D.CL	FR.R	B	Hab, F	○	Most	5-9
Blackcurrant	Ribes nigrum	D.SH	FR	B	Hab, F, P	○	Most.HR	4-8
Blue sausage fruit	Decaisnea fargesii	D.SH	FR		Hab, F	○⊕	ML	4-8
Blueberry	Vaccinium corymbosum	D.SH	FR	B	Hab, F	○⊕	A	3-8
Blueberry climber	Ampelopsis brevipedunculata	D.CL	L.YO	B	Hab, F	○	Most	5-8
Box leaf	Azara microphylla	E.SH	FR	B	F	●	MS	8-10
Box thorn	Lycium barbarum	D.SH	FR.YS	B	F	○	WD	6-9
Buck bush	Ceanothus cuneatus	E.SH	SE	B	F	○	LS.LF	6-9
Buffalo berry	Shepherdia argentea	D.SH	FR	B	F, P	○	MD	3-9
Butternut	Juglans cinerea	D.T	SE.SA.O		F	○	DS.LO	3-7
Camellia tea oil	Camellia oleifera	E.SH	O	B	Hab	⊕●	LF	6-9
Canada buffaloberry	Shepherdia canadensis	D.SH	FR	B	Hab, F		WD	2-6

CREATING YOUR PERMACULTURE HEAVEN — 271

Common Name	Botanical Name	Type	Edible Part or Use	Insectary	Wildlife Value	Light	Soil Type	USDA Zone
Carob	*Ceratonia siliqua*	E.T	SE.SP	B	F	○	WD	9-11
Carolina allspice	*Calycanthus floridus*	D.SH	BA.CS	B	F	○	DM.LL	5-10
Cascara sagrada	*Rhamnus purshiana*	D.T	FR	B	Hab	○⊕	Most	4-9
Chaste tree	*Vitex agnus-castus*	D.SH	FR	B	F	○	CH.SL	7-9
Checkerberry	*Gaultheria procumbens*	E.SH	FR.LE	B	F, P	⊕	LF	3-6
Cherry plum	*Prunus cerasifera*	D.T	FR	B	Hab, F	○	WD.L	5-8
Cherry silverberry	*Elaeagnus multiflora*	D.SH	FR	B	Hab, F, P	○	LS	5-9
Chestnut oak	*Quercus prinus*	D.T	SE		Hab, F, P	○	DL	4-8
Chilean bellflower	*Lapageria rosea*	E.CL	FR	B	Hum, F	●	LF	8-11
Chilean guava	*Myrtus luma*	E.SH	FR	B	F	○⊕	Most	9-11
China root	*Smilax china*	D.CL	R.YS.FR	B	F	○⊕	Any	5-9
Chinese catalpa	*Catalpa ovata*	D.T	FL,PO	B			Most	5-9
Chinese cedar	*Toona sinensis*	D.T	LE.YS	B		○	CH	6-11
Chinese dogwood	*Cornus kousa chinensis*	D.T	FR.YL	B	Hab, F	○⊕	Most	5-8
Chinese gooseberry	*Actinidia chinensis*	D.CL	FR	B	F	○	LO	6-9
Chinese indigo	*Indigofera decora*	D.SH	SE	B		○	WD	5-7
Chinese peashrub	*Caragana sinica*	D.SH	FL	B	Hab	○	LiS	5-9

Common Name	Botanical Name	Type	Edible Part or Use	Insectary	Wildlife Value	Light	Soil Type	USDA Zone
Chinese yellow wood	*Maackia amurensis*	D.SH	YL	B		○	Most	4-7
Chusan palm	*Trachycarpus fortunei*	E.T	YFB	B		●	R	8-11
Clethra barbinervis	*Clethra barbinervis*	D.SH	LE	B		⊕	LF.WD. P	4-8
Climbing bittersweet	*Celastrus scandens*	D.CL	BA.YS	B		○⊕	DL.LF	3-8
Common lime	*Tilia x europaea*	D.T	LE.FL.SA	B	Hab	○	Most	3-9
Common thyme	*Thymus vulgaris*	E.SH	LE	B		○	CH.SL	5-11
Cornelian cherry	*Cornus mas*	D.T	FR	B	F	○	Most	4-8
Cottonwood	*Populus fremontii*	D.T	CA.BA		Hab	○	DM	2-9
Cranberry	*Vaccinium macrocarpon*	E.SH	FR	B	F	○⊕	M.A.P	2-7
Creambush/oceanspray	*Holodiscus discolor*	D.SH	FR	B	Hab	○⊕	Most	4-8
Crowberry	*Empetrum nigrum*	E.SH	FR	B	F, P	○⊕	P	3-8
Damson	*Prunus insititia*	D.T	FR	B	Hab, F	○	Most	5-9
Date plum	*Diospyros lotus*	D.T	FR		Hab, F	○	SL.CL.WD	7-9
Devil's club	*Oplopanax horridus*	D.SH	YS		Hab	●	MS.C	4-8
Digger pine	*Pinus sabiniana*	E.T	SE		Hab, F	○	Most	8-10
Dogwood	*Cornus sanguinea*	D.SH	O	B	Hab, F	○⊕	Most	4-8

CREATING YOUR PERMACULTURE HEAVEN 273

Common Name	Botanical Name	Type	Edible Part or Use	Insectary	Wildlife Value	Light	Soil Type	USDA Zone
Douglas fir	*Pseudotsuga menziesii*	E.T	YS		Hab, F, P	○	Most.LF	3-6
Elderberry (black)	*Sambucus nigra*	D.SH	FL.FR	B	Hab, F, P	○ ● ⊕	Most	5-8
False huckleberry	*Menziesia ferruginea*	D.SH	FR	B	F, P	○ ⊕	WD.LF	5-9
Fig	*Ficus carica*	D.T	FR		Hab, F, P	○	WD.L	6-10
French hydrangea	*Hydrangea macrophylla*	D.SH	LE	B		○ ⊕	WD.LO	5-9
Fringe tree	*Chionanthus virginicus*	D.SH	FR	B	Hab, F	○	Most	4-9
Fuchsia	*Fuchsia magellanica*	D.SH	FR	B	Hum	○ ⊕	WD.HR	8-11
Gooseberry	*Ribes uva-crispa*	D.SH	FR	B	Hab, F, P	○ ⊕	Most	4-8
Grape	*Vitis vinifera*	D.CL	LE.SA.FR.O	B	Hab, F	○	CH.CL.S	6-10
Greasewood	*Sarcobatus vermiculatus*	D.SH	SE		F	○	Most	4-8
Hardy hibiscus/ mallow	*Hibiscus syriacus*	D.SH	LE.FL.O	B	Hab, F, Hum	○	Most.WD	5-9
Hawthorn	*Crataegus monogyna*	D.T	LE.FR	B	Hab, F	○ ⊕	Most	4-8
Hazel nut (corkscrew)	*Corylus avellana*	D.SH	SE.O	B	Hab, F	All	Most	4-8
Hercule's club	*Aralia spinosa*	D.T	YL	B	F	○ ⊕	DL	5-9
Himalayan giant blackberry	*Rubus procerus*	D.CL	FR	B	F	○ ⊕	Most	4-8

Common Name	Botanical Name	Type	Edible Part or Use	Insectary	Wildlife Value	Light	Soil Type	USDA Zone
Hobbleberry	*Viburnum lantanoides*	D.SH	FR	B	F, P	●	R	3-7
Honey locust	*Gleditschia triacanthos*	D.T	SP	B	Hab, F, P	○	LO	4-8
Hyssop	*Hyssopus officinalis*	E.SH	LE	B		○	L.D.LS	5-10
Indian cherry	*Rhamnus carolinianus*	D.T	FR	B	Hab, F, P	○⊕	Most	5-9
Indian hawthorn	*Rhaphiolepis indica*	E.SH	FR	B	F	○	F	8-11
Japanese hawthorn	*Rhaphiolepis umbellata*	E.SH	SE	B	F	○	F	7-10
Japanese ivy	*Parthenocissus tricuspidata*	D.CL	SA	B	F	○⊕●	WD.Most	4-8
Japanese raisin tree	*Hovenia dulcis*	D.T	FR	B	Hab	○	SL	5-9
Japanese rose	*Kerria japonica*	D.SH	YL	B	Hum	○⊕	Most	4-9
Japanese wineberry	*Rubus phoenicolasius*	D.SH	FR	B	Hab, F	○⊕	Most	4-8
Judas tree	*Cercis siliquastrum*	D.T	FL.SP	B	Hab, F	○	LS	6-9
Jujube	*Ziziphus jujuba*	D.T	FR	B	Hab, F	○	Most	5-10
Juneberry	*Amelanchier canadensis*	D.SH	FR	B	Hab, F, P	○	MS	4-7
Juniper	*Juniperus communis*	E.SH	FR		F, P	○⊕	Most.L	4-10
Kamchatka bilberry	*Vaccinium praestans*	E.SH	FR	B	F	⊕	A	4-8

Common Name	Botanical Name	Type	Edible Part or Use	Insectary	Wildlife Value	Light	Soil Type	USDA Zone
Kentucky coffee tree	Gymnocladus dioica	D.T	SP		Hab, F	○	DS	3-8
Lavender	Lavandula angustifolia	E.SH	LE	B		○	Most.WD	5-8
Leafy lespedeza	Lespedeza cyrtobotrya	D.SH	YS	B		○	LS	5-9
Lespedeza	Lespedeza bicolor	D.SH	SE,YS,LE	B	Hab, F	○	LS	4-8
Locust	Robinia pseudoacacia	D.T	FL,SE	B	Hab, F, Hum	○	Most	4-9
Loquat	Eriobotrya japonica	E.T	FR	B	Hum, F	○	WD	8-11
Macqui	Aristotelia chilensis	E.SH	FR	B	F	○	WD,CL,SL	7-10
Manna plant	Tamarix gallica	D.SH	MA	B	Hab	○	DL	4-8
Mayflower	Epigaea repens	E.SH	FR	B	F	●	LF,HR	3-8
Medlar	Mespilus germanica	D.T	FR	B	Hab	○	Most	5-8
Mimosa tree	Albizia julibrisin	D.T	LE	B	Hab	○	Most	7-10
Mooseberry	Viburnum edule	D.SH	FR	B	F, P	○⊕	DP	4-8
Mountain pepper	Drimys lanceolata	E.SH	FR			●⊕	Most.WD	7-10
Myrtle	Myrtus communis	E.SH	FR	B	F	○	Most	9-11
Narrowleaf firethorn	Pyracantha angustifolia	E.SH	FR	B	F	○	WD	6-10

Common Name	Botanical Name	Type	Edible Part or Use	Insectary	Wildlife Value	Light	Soil Type	USDA Zone
Olive	Olea europaea	E.T	FR.O		Hab, F	○	Most.WD	8-10
Oregon grape	Mahonia aquifolium	E.SH	FR	B	F	●	Most	4-8
Oso berry	Oemleria cerasiformis	D.SH	FR	B	Hab, F	○⊕	Most	5-9
Pale wolfberry	Lycium pallidum	D.SH	FR	B	F	○	WD	5-9
Palm lily	Yucca gloriosa	E.SH	RO	B	F, Hum	○	WD.DCH.DPE	6-11
Pawpaw	Asimina triloba	D.SH	FR		Hab, F, P	○	LO	5-8
Paper mulberry	Broussonetia papyrifera	D.T	FR.LE	B	F	○	MS	6-11
Parsley-leaved blackberry	Rubus laciniatus	D.SH	FR	B	F, P	○⊕	Most	4-8
Partridge berry	Mitchella repens	E.SH	FR		F, P	●	LF	4-9
Peach	Prunus persica	D.T	FR.O	B	Hab, F	○	WD	5-9
Pear	Pyrus communis	D.T	FR	B		○	WD	4-9
Pearl berry	Margyricarpus pinnatus	E.SH	FR	B	F	○	WD	8-11
Persimmon	Diospyros kaki	D.T	FR		Hab, F	○	WD.LO	7-10
Pignut hickory	Carya glabra	D.T	SE.SA		Hab, F	○	DL	4-9
Pineapple guava	Feijoa sellowiana	E.T	FR.FL	B	F, P	○	Most	8-11
Plum	Prunus domestica	D.T	FR.O	B	Hab, F	○	WD.L	4-9
Pomegranate	Punica granatum	D.T	FR		Hum	○	WD	8-12
Pygmy peashrub	Caragana pygmaea	D.SH	RO	B	Hab, F, Hum	○	LiS	3-7

Common Name	Botanical Name	Type	Edible Part or Use	Insectary	Wildlife Value	Light	Soil Type	USDA Zone
Quince	Cydonia oblonga	D.T	FR	B		○	Most	5-9
Ramanas rose	Rosa rugosa	D.SH	FR.SE.FL	B	F	○	Most	3-9
Raspberry	Rubus idaeus	D.SH	FR	B	Hab, F, P	○	LS	4-8
Red alder	Alnus rubra	D.T	SA.IB.BU		F	○⊕	HDS	6-8
Red currant	Ribes rubrum	D.SH	FR	B	Hab, F, P	○⊕	M	4-9
Redbud	Cercis canadensis	D.T	FL	B	Hab	○	DSL	4-9
Rock maple	Acer glabrum	D.T	YS.IB	B		○	MD	3-8
Rosemary	Rosmarinus officinalis	E.SH	LE	B		○	LS.D	6-11
Rowan	Sorbus aucuparia	D.T	FR	B	F	○⊕	Most	3-6
Sacred bamboo	Nandina domestica	E.SH	FR.YL		Hab, F	○	R	6-9
Sage brush	Artemisia tridentata	E.SH	LE.SE	B	Hab, P	○	WD.LF	4-10
Sassafras	Sassafras albidum	D.T	LE.FR.BA	B	Hab, F, P	●	F.DE	5-9
Sea buckthorn	Hippophae rhamnoides	D.SH	FR	B	Hab, F	○	M	3-7
Shagbark hickory	Carya ovata	D.T	SE.SA	B	Hab, F, P	○	DL	4-8
Shallon	Gaultheria shallon	E.SH	FR	B	F, P, Hum	⊕	LF	6-9
Shellbark hickory	Carya laciniosa	D.T	SE.SA	B	Hab, F, P	○	DL	5-9
Siberian pea shrub	Caragana arborescens	D.SH	SE,PO	B	F, P	○	LiS	2-7
Silver birch	Betula pendula	D.T	LE.IB.SA	B	F	○⊕	WD	2-6
Smooth sumac	Rhus glabra	D.SH	FR	B	F, P	○	Most	3-9

Common Name	Botanical Name	Type	Edible Part or Use	Insectary	Wildlife Value	Light	Soil Type	USDA Zone
Soapweed	Yucca glauca	E.SH	FR.FL	B	Hab, F	○	WD.DCH.DPE	4-10
Sour cherry	Prunus cerasus	D.T	FR.O	B	Hab, F	○	Most	3-7
Spanish bayonet	Yucca baccata	E.SH	FR.SE.FL	B	F	○	WD	6-11
Spice bush	Lindera benzoin	D.SH	BA	B	F	○⊕	LF	4-9
Spoonleaf yucca	Yucca filamentosa	E.SH	FR	B	Hum	○	WD.DCH.DPE	7-10
Stagberry	Viburnum prunifolium	D.T	FR	B	Hab, F	○⊕	DP.M	3-9
Stone pine	Pinus pinea	E.T	SE		Hab, F	○	CH	7-11
Strawberry tree	Arbutus unedo	E.T	FR	B	Hab, F	○⊕	MS	7-11
Sugarberry	Celtis occidentalis	D.T	FR.SE	B	Hab, F	○⊕	Most	2-9
Sugar maple	Acer saccharum	D.T	SA	B	Hab, F	○	DM	4-8
Sweet buckeye	Aesculus flava	D.T	SE	B	Hab, F, Hum	○⊕	DS	4-8
Sweet chestnut	Castanea sativa	D.T	SE	B	Hab, F, P	○	Most.LF	5-7
Sweet fern	Comptonia peregrina	D.SH	FR	B	Hab, F	○⊕	LF.P	4-8
Tanbark oak	Lithocarpus densiflorus	E.T	SE	B	Hab, F	●	Most.WD	6-9
Tara vine	Actinidia arguta	D.CL	FR.SA	B		○	LO	4-8
Tatarian maple	Acer tataricum	D.T	SE.SA	B		○	MD	3-7
Tulip tree	Liriodendron tulipifera	D.T	BA	B	F, Hum	●	DS	4-9
Tupelo	Nyssa sylvatica	D.T	FR	B	Hab, F, P	○⊕	LF	4-9

Common Name	Botanical Name	Type	Edible Part or Use	Insectary	Wildlife Value	Light	Soil Type	USDA Zone
Varnish tree	Koelreuteria paniculata	D.T	LE.FR	B	Hab, F	○	Most	5-8
Violet willow	Salix daphnoides	D.T	YS.IB	B		○⊕	Most	4-8
Virginia creeper	Parthenocissus quinquefolia	D.CL	FR	B	F	○⊕●	Most.WD	3-10
Walnut	Juglans regia	D.T	SE.SA.O	B	Hab, F	○	Most.WD.DS	7-9
Weeping forsythia	Forsythia suspensa	D.SH	YL	B	F	○⊕	Most	5-8
Western hemlock	Tsuga heterophylla	E.T	IB		Hab	○⊕●	DCH	6-7
White mulberry	Morus alba	D.T	FR.YL		Hab, F, P	○	WD	4-9
White willow	Salix alba	D.T	LE.IB	B	Hab	○⊕	R	2-8
Whitebark pine	Pinus albicaulis	E.T	SE		Hab, F		WD	4-8
Winter savoury	Satureia montana	E.SH	LE	B		○	WD	6-11
Winter's bark	Drimys winteri	E.SH	BA		Hab	⊕●	M	7-10
Witch hazel	Hamamelis virginiana	D.SH	SE	B	Hab, F, P	○⊕	MS	3-8
Worcesterberry	Ribes divaricatum	D.SH	FR	B	F	○⊕	Most	4-8
Yew	Taxus baccata	E.T	FR		Hab, F	●	Most	5-7
Yellowhorn	Xanthoceras sorbifolium	D.T	LE.FL.FR	B	F	○	CH.SL.WD	4-7

Table 2—plants suitable for ground cover

Type of plant
A = Annual
B = Bamboo
Bi = Biennial
Cl—Climber
F = Fern
Sh = Shrub
T = Tree
P = Perennial

Deciduous/Evergreen
D = Deciduous
E = Evergreen

Soil
L = Light
M = Medium
H = Heavy

Shade
○—prefers sun
⊕—prefers partial shade
●—prefers shade

Moisture
D = Succeeds in dry soils
M = Succeeds in moist soils (the average soil moisture level)
We = Succeeds in wet soils
Wa = Succeeds in water

pH
A = Succeeds in acid soils
N = Succeeds in neutral soils
B = Succeeds in basic (alkaline soils)

Growth rate
S = Slow
M = Medium
F = Fast

CREATING YOUR PERMACULTURE HEAVEN 281

Botanical Name	Common Name	Habit	Evergreen/ Deciduous	Height (meters)	US Hardiness Zone	Soil	Shade	Moisture	pH	Growth Rate	Notes
Acaenia anserinifolia	Pirri-pirri bur	P	E	0.10	5-9	LMH	○⊕	M	ANB	F	Leaves used for tea.
Acanthus mollis	Bear's breeches	P	E	1.20	6-10	LM	○⊕	DM	ANB	S	Drought tolerant.
Adiantum venustum	Evergreen maidenhair fern	F	E	0.25	9-11	LMH	⊕	M	ANB	M	
Aegopodium podagraria	Ground elder	P	D	0.60	4-9	LMH	○⊕●	M	ANB	F	Edible leaves.
Ajuga australis	Australian bugle	P	E	0.15	5-9	LMH	○	DM	ANB	F	
Ajuga reptans	Bugle	P	E	0.30	3-10	LMH	○⊕●	DMWe	ANB	M	Edible young shoots. Drought tolerant.
Alchemilla xanthochlora	Lady's mantle	P	D	0.30	4-7	LMH	○⊕	DM	NB	S	Edible young leaves.
Antennaria dioica	Catsfoot	P	E	0.12	4-8	L	○	DM	ANB	S	Tolerates poor soil and drought.
Arabis caucasica	Rock cress	P	E	0.15	4-9	LMH	○⊕	DM	ANB	M	Edible leaves. Tolerates poor soil and drought.
Arctostaphylos nevadensis	Pine-mat manzanita	Sh	E	0.10	5-9	LM	○⊕	M	A	M	Edible fruit and seed. Dye.
Arctostaphylos uva-ursi	Bearberry	Sh	E	0.10	4-8	LM	○⊕●	M	ANB	M	Edible fruit. Dye.
Arisarum vulgare	Friar's cowl	P	E	0.45	6-9	LM	⊕●	M	ANB		Edible root if thoroughly cooked.
Armeria maritima	Thrift	P	E	0.10	3-9	LM	○	DM	ANB	M	Edible leaves. Tolerates maritime exposure.
Artemisia ludoviciana	White sage	P	D	1.00	4-9	LM	○⊕	DM	NB	M	Condiment, leaves, tea. Drought tolerant.

Botanical Name	Common Name	Habit	Evergreen/ Deciduous	Height (meters)	US Hardiness Zone	Soil	Shade	Moisture	pH	Growth Rate	Notes
Artemisia stelleriana	Beach wormwood	P	E	0.50	3-9	LM	○	DM	ANB	M	Condiment, tea. Medicinal. Drought tolerant.
Asarum canadense	Snake root	P	D	0.30	3-9	LMH	⊕	M	ANB	S	Root is a ginger substitute. Medicinal.
Asarum caudatum	Wild ginger	P	E	0.10	6-10	LMH	⊕	M	AN	S	Root is a ginger substitute.
Asarum europaeum	Asarabacca	P	E	0.10	4-8	LMH	⊕	M	ANB	S	Dye. Medicinal.
Asarum shuttleworthii	Mottled wild ginger	P	E	0.10	5-9	LMH	⊕	M	AN	S	Root is a ginger substitute.
Aspidistra elatior	Aspidistra	P	E	0.60	7-11	LMH	⊕	DM	ANB	S	Medicinal. Tolerates drought and poor soils.
Asplenium scolopendrium	Hart's tongue Fern	F	E	0.60	4-8	LM	●	DM	ANB	M	Medicinal.
Athyrium filix-femina	Common Lady Fern	F	D	0.60	3-8	LMH	⊕	M	ANB	M	Edible young shoots and root—they must be well cooked.
Aubrieta deltoidea	Aubretia	P	E	0.15	4-9	LM	○⊕	DM	ANB	M	Drought tolerant.
Aurinia saxatilis	Golden alyssum	P	E	0.25	4-10	LMH	○	DM	ANB	M	Tolerates drought and poor soils.
Baccharis pilularis	Chaparral broom	Sh	E	0.50	8-10	LMH	○	DM	ANB	M	Tolerates poor soils. A good ground stabilizer.

Botanical Name	Common Name	Habit	Evergreen/ Deciduous	Height (meters)	US Hardiness Zone	Soil	Shade	Moisture	pH	Growth Rate	Notes
Baccharis viminea	Mule fat	Sh	E	4.00	7-10	LMH	○	DM	ANB	M	Tolerates poor soils. A good ground stabilizer.
Ballota pseudodictamnus	False dittany	P	E	0.60	7-10	LMH	○	DM	ANB	M	Flowers are used as candle wicks.
Bergenia ciliata	Fringed bergenia	P	E	0.30	6-9	LMH	○⊕	M	ANB	M	Medicinal. Tolerates poor soils.
Bergenia crassifolia	Siberian tea	P	E	0.30	3-7	LMH	○⊕	M	ANB	S	A tea is made from the leaves.
Bergenia purpurascens	Purple bergenia	P	E	0.45	4-8	LMH	○⊕	M	ANB	S	Medicinal.
Blechnum spicant	Hard Fern	F	E	0.30	4-8	LMH	○⊕●	M	AN	S	Root and young leaves—cooked.
Brachyglottis 'Sunshine'	Ragwort sunshine	Sh	E	1.50	7-10	LMH	○	M	ANB	S	Tolerates maritime exposure and drought.
Calluna vulgaris	Heather	Sh	E	0.60	4-7	LM	○⊕	DM	A	M	Tea, dye, thatching, basketry, brooms.
Campanula portenschlagiana	Adria bellflower	P	E	0.25	3-7	LM	○⊕	M	NB	F	Edible leaves.
Campanula poscharskyana	Trailing bellflower	P	E	0.25	3-7	LM	○⊕	M	NB	F	Edible leaves.
Campsis radicans	Trumpet vine	Cl	D	12.00	4-10	LMH	○⊕	M	ANB	F	Medicinal.
Carex elata	Tufted sedge	P	E	1.20	5-9	LMH	○⊕	MWe	ANB	M	Edible root and seed.
Ceanothus prostratus	Squaw carpet	Sh	E	0.05	6-9	LM	○⊕	DM	ANB	M	Dye, soap substitute.

Botanical Name	Common Name	Habit	Evergreen/Deciduous	Height (meters)	US Hardiness Zone	Soil	Shade	Moisture	pH	Growth Rate	Notes
Ceanothus purpureus	Hollyleaf ceanothus	Sh	E	2.00	7-10	LM	○ ⊕	DM	ANB	M	Dye, soap substitute.
Ceanothus thyrsiflorus	Blue brush	Sh	E	4.50	7-9	LM	○ ⊕	DM	ANB	M	Dye, soap substitute. Tolerates maritime exposure.
Centaurea montana	Mountain cornflower	P	D	0.45	3-9	LMH	○	DM	ANB	F	Succeeds in poor soils.
Cephalotaxus fortunei	Chinese plum yew	Sh	E	6.00	6-9	LMH	○ ⊕ ●	M	ANB	S	Edible seed and fruit.
Chaenomeles speciosa	Japanese quince	Sh	D	3.00	4-8	LMH	○ ⊕ ●	M	ANB	M	Edible fruit.
Chaenomeles x superba	Dwarf quince	Sh	D	1.00	5-8	LMH	○ ⊕ ●	M	ANB	F	Edible fruit.
Chamaemelum nobile	Chamomile	P	E	0.15	4-8	LMH	○ ⊕	DM	ANB	M	Tea, condiment, hair shampoo, dye, companion plant, liquid feed and plant tonic, medicinal.
Chelidonium majus	Greater celandine	P	E	0.50	5-8	LMH	○ ⊕ ●	M	ANB	F	Medicinal. A weed in the garden, only use it in wild areas.
Chrysosplenium alternifolium	Golden saxifrage	P	E	0.30	4-8	LMH	○ ⊕	MWe	ANB	M	Edible leaves.
Chrysosplenium americanum	Water mat	P	E	0.10	3-7	LMH	○ ⊕	MWe	ANB	M	Condiment.
Chrysosplenium oppositifolium	Golden saxifrage	P	E	0.15	4-8	LMH	○ ⊕	MWe	ANB	M	Edible leaves.
Cistus salviifolius	Rock rose	Sh	E	0.60	8-11	LM	○	DM	ANB	M	Condiment. Tolerates maritime exposure.

CREATING YOUR PERMACULTURE HEAVEN   285

Botanical Name	Common Name	Habit	Evergreen/ Deciduous	Height (meters)	US Hardiness Zone	Soil	Shade	Moisture	pH	Growth Rate	Notes
Clematis orientalis	Oriental virginsbower	Cl	D	2.00	5-9	LMH	○ ⊕	M	ANB	F	Medicinal.
Convallaria keiskei	Lily of the valley	P	D	0.30	3-7	LMH	○ ⊕ ●	DMWe	ANB	S	Medicinal.
Convallaria majalis	European Lily of the valley	P	D	0.15	2-7	LMH	○ ⊕ ●	DMWe	ANB	S	Essential oil, dye. Medicinal.
Coprosma acerosa	Sand Coprosma	Sh	E	0.10	7-10	LM	○ ⊕	M	AN	S	Edible fruit. Coffee substitute. Dye.
Coptis chinensis	Huang Lian	P	E	0.20	5-9	LMH	○ ⊕	MWe	A		Medicinal.
Coptis japonica		P	E	0.22	6-9	LMH	○ ⊕	M	A		Medicinal.
Coptis occidentalis	Idaho goldthread	P	E	0.15	4-8	LMH	○ ⊕	M	A		Dye. Medicinal.
Coptis trifolia	Goldthread	P	E	0.15	2-7	LMH	⊕	M	A		Dye. Medicinal.
Coriaria microphylla		Sh	D	1.20	7-10	LMH	○ ⊕	M	ANB	M	Dye, ink, tannin.
Cornus canadensis	Creeping dogwood	P	D	0.25	2-7	LMH	○ ⊕	M	AN	M	Edible fruit.
Cornus suecica	Dwarf cornel	P	D	0.15	2-6	LM	○ ⊕	M	AN	M	Edible fruit.
Coronilla varia	Crown vetch	P	D	0.30	4-9	LMH	○	DM	ANB	M	Insecticide. Soil stabilizer. Tolerates maritime exposure and poor soils.
Cotoneaster conspicuus	Tibetan cotoneaster	Sh	E	3.00	6-8	LMH	○ ⊕	M	ANB	M	Dye. Wind tolerant.
Cotoneaster microphyllus	Littleleaf cotoneaster	Sh	E	1.00	4-8	LMH	○ ⊕	M	ANB	S	Dye. Wind tolerant.
Cyathodes colensoi		Sh	E	0.45	7-10	LMH	⊕	M	AN	S	Edible fruit.

Botanical Name	Common Name	Habit	Evergreen/ Deciduous	Height (meters)	US Hardiness Zone	Soil	Shade	Moisture	pH	Growth Rate	Notes
Cyathodes fraseri		Sh	E	0.15	7-10	LMH	●	M	AN	S	Edible fruit.
Cystopteris bulbifera	Berry bladder fern	F	D	0.9	4-8	LMH	⊕●	M	ANB	F	Edible root. The plant can become invasive.
Cytisus scoparius	Broom	Sh	D	2.40	5-8	LMH	○⊕	DM	ANB	F	Succeeds on poor soils. Condiment, coffee substitute, fibre, dye, brooms. Medicinal.
Dactylis glomerata	Cock's foot	P	E	1.00	4-8	LMH	○⊕	M	ANB	F	A food plant for native wildlife.
Darmera peltata	Umbrella plant	P	D	2.00	5-8	LMH	○⊕	WeWa	ANB	S	Edible leaf stalk. Soil stabilization.
Deschampsia caespitosa	Tussock grass	P	E	1.50	5-9	LMH	○⊕	MWe	ANB	M	Edible seed.
Dianthus gratianopolitanus	Cheddar pink	P	E	0.30	3-7	LMH	○	D	NB	M	
Dryas octopetala	Mountain avens	Sh	E	0.10	3-6	LMH	○	M	ANB	M	Tea substitute.
Dryopteris filix-mas	Male Fern	F	D	1.20	6-9	LMH	⊕	DM	AN	M	Edible young fronds—must be cooked.
Duchesnea indica	Indian strawberry	P	E	0.10	5-11	LMH	○	M	ANB	F	Edible fruit.
Elymus arenarius	Lyme grass	P	E	1.20	5-9	LMH	○	DM	ANB	F	Weaving, soil stabilization. Tolerates maritime exposure.
Empetrum atropurpureum	Purple crowberry	Sh	E	0.20	2-7	LM	○⊕	M	A	S	Edible fruit. Wind tolerant.

Botanical Name	Common Name	Habit	Evergreen/ Deciduous	Height (meters)	US Hardiness Zone	Soil	Shade	Moisture	pH	Growth Rate	Notes
Empetrum eamesii	Rockberry	Sh	E	0.15	3-8	LM	○ ⊕	M	A	S	Edible fruit. Wind tolerant.
Empetrum eamesii hermaphroditum	Mountain crowberry	Sh	E	0.30	3-8	LMH	○	M	AN	S	Edible fruit. Wind tolerant.
Empetrum nigrum	Crowberry	Sh	E	0.30	3-8	LMH	○ ⊕	M	AN	S	Edible fruit. Dye. Wind tolerant.
Empetrum rubrum	Crowberry	Sh	E	0.30	2-7	LMH	○ ⊕	M	AN	S	Edible fruit. Wind tolerant.
Epigaea asiatica		Sh	E	0.10	4-8	LM	⊕ ●	M	AN		Edible fruit.
Epimedium grandiflorum	Buckland spider	P	D	0.35	4-8	LMH	⊕	M	ANB	M	Young leaves—cooked.
Epimedium sagittatum	Yin yang huo	P	D	0.50	5-9	LMH	⊕	M	ANB	M	Young leaves—cooked.
Erica vagans	Cornish heath	Sh	E	0.75	4-8	LM	○	DM	AN	M	Dye, brooms, thatching.
Erica x darleyensis	Cape heath	Sh	E	0.60	6-9	LM	○	M	ANB	M	Dye, brooms, thatching.
Euonymus fortunei	Winter creeper	Sh	E	4.50	5-9	LMH	○ ⊕ ●	DM	ANB	M	Plants can also be grown as a hedge
Fragaria californica	Californian strawberry	P	E	0.30	5-9	LMH	○ ⊕	M	ANB	F	Edible fruit.
Fragaria daltoniana		P	E	0.30	5-9	LMH	○ ⊕	M	ANB	F	Edible fruit.
Fragaria moschata	Hautbois strawberry	P	E	0.45	5-9	LMH	○ ⊕	M	ANB	F	Edible fruit.
Fragaria viridis	Green strawberry	P	E	0.30	5-9	LMH	○ ⊕	M	ANB	F	Edible fruit.
Galium mollugo	Hedge bedstraw	P	D	1.20	3-7	LMH	⊕	DM	ANB	M	Edible leaves. Dye.
Galium odoratum	Sweet woodruff	P	D	0.15	5-9	LMH	⊕ ●	DM	ANB	M	Edible leaves, tea. Dye. Medicinal.

Botanical Name	Common Name	Habit	Evergreen/ Deciduous	Height (meters)	US Hardiness Zone	Soil	Shade	Moisture	pH	Growth Rate	Notes
Gaultheria adenothrix		Sh	E	0.30	8-11	LM	◉	M	AN	M	Edible fruit.
Gaultheria depressa	Mountain snowberry	Sh	E	0.10	8-11	LM	○⊕	M	AN	M	Edible fruit.
Gaultheria hispidula	Creeping snowberry	Sh	E	0.10	5-9	LM	⊕	M	AN	F	Edible fruit.
Gaultheria humifusa	Alpine wintergreen	Sh	E	0.10	6-9	LM	○⊕	M	AN		Edible fruit.
Gaultheria macrostigma		Sh	E	0.50	7-10	LM	⊕	M	A		Edible fruit.
Gaultheria mucronata	Prickly heath	Sh	E	1.50	5-9	LM	○⊕	M	A		Edible fruit.
Gaultheria myrsinoides		Sh	E	0.15	8-11	LM	○⊕	M	A		Edible fruit.
Gaultheria nummularioides		Sh	E	0.10	8-11	LM	○⊕	M	AN		Edible fruit.
Gaultheria ovatifolia	Mountain Checkerberry	Sh	E	0.15	5-9	LM	⊕	M	AN	M	Edible fruit.
Gaultheria procumbens	Checkerberry	Sh	E	0.15	3-6	LM	⊕	DM	AN	M	Edible fruit and leaves. Tea. Essential oil, Medicinal.
Gaultheria pumila leucocarpa		Sh	E	0.20	6-9	LM	○⊕	M	A	F	Edible fruit.
Gaultheria shallon	Shallon	Sh	E	1.20	6-9	LM	⊕	DM	A	M	Edible fruit. Tea. Dye.
Gaultheria trichophylla		Sh	E	0.10	7-10	LM	○⊕	M	AN		Edible fruit.
Gaultheria x wisleyensis	Hybrid wintergreen	Sh	E	1.00	5-9	LM	⊕	M	A		Edible fruit.
Gaylussacia brachycera	Box huckleberry	Sh	E	0.45	5-9	LMH	○⊕	DM	A		Edible fruit.
Genista hispanica	Spanish gorse	Sh	D	0.50	5-9	LM	○	DM	ANB		Medicinal. Drought tolerant.

Botanical Name	Common Name	Habit	Evergreen/ Deciduous	Height (meters)	US Hardiness Zone	Soil	Shade	Moisture	pH	Growth Rate	Notes
Geranium macrorrhizum	Bigroot Geranium	P	D	0.50	4-8	LMH	○ ⊕	DM	NB	F	Essential oil.
Geranium sylvaticum	Wood cranesbill	P	D	0.60	4-8	LMH	○ ⊕ ●	M	ANB	F	Essential oil.
Geranium wallichianum	Geranium	P	D	0.30	6-9	LMH	○ ⊕	M	ANB	F	Dye.
Glechoma hederacea	Ground ivy	P	E	0.20	3-10	MH	⊕	M	ANB	F	Edible leaves. Tea.
Hebe 'Great Orme'	Great Orme Hebe	Sh	E	1.00	5-9	LMH	○	M	ANB	M	Can also be used as a hedge. Tolerates maritime exposure.
Hebe rakaiensis	Rakai hebe	Sh	E	1.00	5-9	LMH	○	M	ANB	M	Can also be grown as a hedge.
Hedera helix	Ivy	Cl	E	15.00	5-11	LMH	○ ⊕ ●	MWe	ANB	M	Dye, hair shampoo. An excellent plant for wildlife.
Herniaria glabra	Rupture wort	Bi/P	E	0.02	4-8	LM	○	DM	ANB	S	Medicinal.
Heuchera americana	Rock geranium	P	E	0.45	5-9	LM	○ ⊕	M	ANB		Mordant.
Heuchera cylindrica	Alum root	P	E	0.45	4-8	LM	○ ⊕	M	ANB		Edible leaves. Mordant.
Heuchera diversifolia	Alum root	P	E	0.45	4-8	LM	○ ⊕	M	ANB		Edible leaves. Mordant.
Heuchera glabra	Alpine Heuchera	P	E	0.60	4-8	LM	○ ⊕	M	ANB		Edible leaves. Mordant.
Heuchera micrantha	Alum root	P	E	0.60	4-8	LM	○ ⊕	M	ANB		Edible leaves. Mordant.
Heuchera sanguinea	Alum root	P	E	0.60	4-10	LM	○ ⊕	M	ANB		Mordant.

Botanical Name	Common Name	Habit	Evergreen/ Deciduous	Height (meters)	US Hardiness Zone	Soil	Shade	Moisture	pH	Growth Rate	Notes
Heuchera versicolor	Pink alumroot	P	E	0.20	8-11	LM	○⊕	M	ANB		Mordant.
Hosta crispula	Curled plantain lily	P	D	0.30	4-8	LMH	○⊕	M	ANB	M	Edible leaves and stems.
Hosta longipes	Small rock plantain lily	P	D	0.30	4-8	LMH	○⊕●	M	ANB	M	Edible leaves and stems.
Hosta plantaginae	August Lily	P	D	0.25	4-8	LMH	○⊕●	M	ANB	M	Edible leaves and stems.
Hosta rectifolia		P	D	0.20	4-8	LMH	○⊕●	M	ANB	M	Edible leaves and stems.
Hosta sieboldiana	Hosta 'big daddy'	P	D	0.60	4-8	LMH	○⊕●	M	ANB	M	Edible leaves and stems.
Hosta undulata	Variegated wavy plantain lily	P	D	0.30	4-8	LMH	○⊕●	M	ANB	M	Edible leaves and stems.
Hosta ventricosa	Blue plantain lily	P	D	0.25	3-9	LMH	○⊕●	M	ANB	M	Edible leaves and stems.
Houttuynia cordata	Tsi	P	D	0.60	5-10	LMH	⊕	MWeWa	ANB	F	Edible leaves.
Hydrangea anomala	Hydrangea	Cl	D	12.00	4-8	LMH	○⊕●	M	ANB	M	Edible leaves. Sugar. Paper. Wind tolerant.
Hypericum androsaemum	Tutsan	Sh	D	1.00	5-10	LMH	○⊕	DM	ANB	M	Medicinal. Wind tolerant.
Hypericum bellum	Beautiful St John's wort	Sh	E	0.70	5-9	LMH	○⊕	M	ANB		
Hypericum calycinum	Rose of Sharon	Sh	E	0.30	5-10	LMH	○⊕●	DM	ANB	F	Dye.
Imperata cylindrica	Cogongrass	P	E	1.20	6-9	LMH	○	DM	ANB	F	Edible young shoots and root. Weaving, paper making, thatching. Tolerates maritime exposure.

Botanical Name	Common Name	Habit	Evergreen/ Deciduous	Height (meters)	US Hardiness Zone	Soil	Shade	Moisture	pH	Growth Rate	Notes
Iris foetidissima	Stinking gladwin	P	E	0.60	7-10	LMH	○ ⊕ ●	DM	ANB	S	Medicinal. Tolerates maritime exposure.
Juniperus communis	Juniper	Sh	E	9.00	4-10	LMH	○ ⊕	DM	ANB	S	Edible fruit, condiment, coffee, tea. Essential oil, incense, insect repellent. Medicinal.
Juniper communis nana	Juniper	Sh	E	9.00	4-10	LMH	○ ⊕	DM	ANB	S	Same as above.
Juniperus conferta	Shore juniper	Sh	E	0.5	6-10	LMH	○ ⊕	DM	ANB	S	Edible fruit. Tolerates maritime exposure.
Juniperus horizontalis	Creeping juniper	Sh	E	1.00	4-9	LMH	○	DM	ANB	S	Coffee, tea. Tolerates maritime exposure.
Juniperus sabina	Savine	Sh	E	4.00	4-7	LMH	○	DM	ANB	S	Essential oil, parasiticide, insect repellent.
Juniperus squamata	Flaky juniper	Sh	E	8.00	4-7	LMH	○	DM	ANB	S	Medicinal. Drought tolerant.
Lamium album	White dead nettle	P	D	0.60	5-9	LMH	○ ⊕	M	ANB	M	Edible young leaves.
Lamium galeobdolon	Yellow archangel	P	D	0.30	3-9	LMH	○ ⊕ ●	M	ANB	F	Edible young leaves and flowers.
Lathyrus latifolius	Perennial sweet pea	P Cl	D	2.00	5-9	LMH	○ ⊕	DM	ANB	F	Seed, seedpods and young plants might be edible. Drought tolerant.

Botanical Name	Common Name	Habit	Evergreen/ Deciduous	Height (meters)	US Hardiness Zone	Soil	Shade	Moisture	pH	Growth Rate	Notes
Liriope graminifolia	Lilyturf	P	E	0.30	7-10	LM	○⊕	DM	ANB	M	Edible root. Drought tolerant.
Liriope minor		P	E	0.30	7-10	LM	○⊕	DM	ANB	M	Edible root. Drought tolerant.
Liriope muscari	Big blue lilyturf	P	E	0.45	5-10	LM	○⊕●	DM	ANB	M	Edible root. Drought tolerant.
Liriope spicata	Creeping Lilyturf	P	E	0.30	4-10	LM	○⊕	DM	ANB	M	Edible root. Drought tolerant.
Lonicera henryi	Copper beauty	Cl	E	4.00	4-8	LMH	○⊕	M	ANB	F	Edible flowers, leaves and stems.
Lonicera japonica	Japanese Honeysuckle	Cl	E	5.00	4-10	LMH	○⊕	DM	ANB	F	Edible flowers. Tea.
Lonicera nitida	Boxleaf honeysuckle	Sh	E	3.60	6-9	LMH	○⊕●	M	ANB	F	Can also be used as a hedge.
Lonicera pileata	Privet honeysuckle	Sh	E	0.20	4-8	LMH	○⊕●	DM	ANB	M	Tolerates maritime exposure.
Lysimachia nummularia	Creeping Jenny	P	E	0.15	4-8	LMH	○⊕	MWeWa	ANB	M	Tea.
Mahonia aquifolium	Oregon grape	Sh	E	2.00	4-8	LMH	○⊕●	DM	ANB	M	Edible fruit and flowers. Dye.
Mahonia repens	Oregon grape	Sh	E	0.30	4-8	LMH	○⊕	DM	ANB	M	Edible fruit. Dye.
Maianthemum canadense	Canada Beadruby	P	D	0.20	3-7	LMH	⊕●	M	ANB	M	The fruit might be edible.
Matteuccia struthiopteris	Ostrich Fern	F	D	0.90	2-7	LMH	⊕	M	AN	M	Edible young leaves and root.

Botanical Name	Common Name	Habit	Evergreen/ Deciduous	Height (meters)	US Hardiness Zone	Soil	Shade	Moisture	pH	Growth Rate	Notes
Meehania urticifolia	Japanese dead nettle	P	D	0.30	4-8	LMH	⊕	M	ANB		Edible young leaves.
Mentha requienii	Corsican mint	P	D	0.10	5-9	LMH	○⊕	M	ANB	S	Edible young leaves. Tea.
Milium effusum	Wood millet	P	E	1.80	5-9	LMH	⊕●	DM	ANB	M	Edible seed. Weaving, soil stabilization.
Miscanthus sinensis	Eulalia	P	D	4.00	4-9	LMH	○⊕	M	ANB	F	Biomass. Wind tolerant.
Mitchella repens	Partridge berry	Sh	E	0.30	4-9	LM	○⊕	M	AN		Edible fruit.
Mitchella undulata		P	E	0.10	5-9	LM	○⊕	M	AN		Edible fruit.
Montia sibirica	Pink purslane (Claytonia sibirica)	A/P	E	0.15	3-7	LMH	○⊕●	DM	ANB	M	Edible leaves and flowers.
Muehlenbeckia axillaris	Sprawling wirevine	Cl	D	0.30	7-10	LMH	○⊕	M	ANB	M	Edible fruit.
Oenanthe javanica	Water dropwort	P	D	1.00	5-11	LMH	○	WeWa	ANB		Edible root, leaves and seed.
Onoclea sensibilis	Sensitive Fern	F	D	1.20	4-8	LMH	⊕●	MWe	AN		Edible young leaves and root.
Ophiopogon japonicus	Dwarf lilyturf	P	E	0.30	7-11	LM	○⊕	MWeWa	ANB		Edible root.
Origanum vulgare	Oregano	P	D	0.60	4-10	LMH	○⊕	DM	ANB	M	Edible leaves, tea, condiment. Essential oil, dye, insect repellent. Medicinal.

Botanical Name	Common Name	Habit	Evergreen/ Deciduous	Height (meters)	US Hardiness Zone	Soil	Shade	Moisture	pH	Growth Rate	Notes
Origanum vulgare hirtum	Greek oregano	P	D	0.60	4-8	LMH	○ ⊕	DM	ANB	M	As above.
Oxalis oregana	Redwood sorrel	P	E	0.20	6-9	LMH	○ ⊕ ●	M	ANB	M	Edible young leaves.
Pachyphragma macrophylla	Thlaspi macrophyllum	P	E	0.40	6-9	LMH	⊕ ●	DM	ANB		Edible leaves?
Pachysandra terminalis	Japanese Spurge	Sh	E	0.20	4-8	LMH	⊕ ●	DM	ANB	M	Edible fruit. Drought tolerant.
Paronychia argentea	Algerian tea	P	E	0.10	6-9	LM	○	DM	ANB		Tea substitute.
Parthenocissus quinquefolia	Virginia creeper	Cl	D	30.00	3-10	LMH	○ ⊕	M	ANB	F	Edible fruit and stems. Dye.
Parthenocissus tricuspidata	Boston ivy	Cl	D	18.00	4-8	LMH	○ ⊕	M	ANB	F	Edible sap.
Peltaria alliacea	Garlic cress	P	E	0.30	5-9	LM	○ ⊕	M	ANB	M	Edible leaves.
Pennisetum alopecuroides	Chinese Fountain Grass	P	D	1.50	5-9	LM	○	DM	ANB		Edible seed.
Petasites albus	Butterbur	P	D	0.60	4-8	LMH	○ ⊕ ●	M	ANB	F	Edible leaf stems.
Petasites frigidus	Sweet coltsfoot	P	D	0.15	6-10	LMH	○ ⊕ ●	MWe	ANB	F	Edible leaves, stems and roots.

Botanical Name	Common Name	Habit	Evergreen/ Deciduous	Height (meters)	US Hardiness Zone	Soil	Shade	Moisture	pH	Growth Rate	Notes
Petasites japonicus	Sweet coltsfoot	P	D	1.50	5-9	LMH	○ ●	MWe	ANB	F	Edible leaf stems and flowers.
Phalaris arundinacea	Canary grass	P	E	1.50	4-9	LMH	○ ⊕	DMWe	ANB	F	
Phlomis fruticosa	Jerusalem sage	Sh	E	1.30	4-7	LMH	○	DM	ANB	M	Succeeds in poor soils.
Phlomis russeliana	Jerusalem Sage	P	D	1.00	5-9	LMH	○ ⊕	DM	ANB	F	Drought tolerant.
Phlomis samia	Turkish sage	P	D	1.00	6-9	LMH	○ ⊕	DM	ANB	F	
Phyla nodiflora	Frogfruit	P	E	0.20	8-12	LMH	○	DM	ANB		Edible leaves. Tea.
Picea abies	Norway spruce	T	E	60.00	2-7	LMH	○	MWe	AN	F	Selected dwarf cultivars are used. Edible cones, inner bark, tea. Adhesive, turpentine.
Pinus mugo	Dwarf mountain pine	T	E	4.50	2-7	LM	○	DM	AN	M	Condiment, dye. Soil stabilization.
Platycrater arguta		Sh	D	0.20	7-10	LMH	⊕	M	ANB	F	Tea. Wind tolerant.
Pleioblastus humilis	Sasa pumila	B	E	2.00	6-9	LMH	⊕	M	ANB	F	Soil stabilization.
Pleioblastus humilis pumilus	Dwarf bamboo	B	E	1.00	6-9	LMH	⊕	M	ANB	F	Soil stabilization.
Pleioblastus pygmaeus	Fern-leaf bamboo	B	E	0.25	7-10	LMH	○ ⊕	M	ANB	F	Soil stabilization.

Botanical Name	Common Name	Habit	Evergreen/ Deciduous	Height (meters)	US Hardiness Zone	Soil	Shade	Moisture	pH	Growth Rate	Notes
Pleioblastus pygmaeus distichus	Pigmy Bamboo	B	E	0.75	7-10	LMH	○ ⊕	M	ANB	F	Soil stabilization.
Pleioblastus variegatus	White-striped bamboo	B	E	0.75	7-10	LMH	○ ⊕	M	ANB	F	Soil stabilization.
Polypodium vulgare	Polypody	F	E	0.30	3-6	LMH	⊕	DM	ANB	M	Edible root.
Potentilla fruticosa	Shrubby cinquefoil	Sh	D	1.20	2-6	LMH	○ ⊕	M	ANB	M	Tea. Drought tolerant.
Primula vulgaris	Primrose	P	E	0.30	5-10	LMH	○ ⊕	M	ANB	M	Edible young leaves and flowers.
Prunella grandiflora	Prunella x webboana	P	E	0.15	4-8	LMH	○ ⊕	M	ANB	M	Edible leaves.
Prunella vulgaris	Self-heal	P	E	0.15	3-7	LMH	○ ⊕	M	ANB	M	Edible leaves, drink. Dye.
Prunus laurocerasus	Cherry laurel	Sh	E	6.00	6-8	LMH	○ ⊕ ●	M	ANB	M	Edible fruit. Dye. Use selected cultivars.
Pueraria lobata	Kudzu vine	P Cl	D	15.00	5-9	LMH	○	M	ANB	F	Edible root, stem and leaves. Fibre.
Pulmonaria officinalis	Lungwort	P	E	0.30	6-9	LMH	⊕ ●	M	ANB	M	Edible leaves.

Botanical Name	Common Name	Habit	Evergreen/ Deciduous	Height (meters)	US Hardiness Zone	Soil	Shade	Moisture	pH	Growth Rate	Notes
Pulmonaria saccharata	Jerusalem sage	P	E	0.30	4-8	LMH	○ ⊕ ●	M	ANB	M	Condiment.
Reineckia carnea		P	E	0.10	6-9	LMH	⊕ ●	M	AN		
Rheum palmatum	Turkey rhubarb	P	D	3.00	6-9	MH	○ ⊕	M	ANB	M	Edible leaf stem. Insecticide.
Ribes glandulosum	Skunk currant	Sh	D	0.40	6-9	LMH	○ ⊕	M	ANB		Edible fruit.
Rosa wichuraiana	Memorial Rosa	Sh	D	0.30	5-8	LMH	○ ⊕	M	ANB	M	Edible fruit and young shoots.
Rosmarinus officinalis	Rosemary	Sh	E	2.00	6-11	LM	○	DM	ANB	M	Condiment, tea. Insect repellent, incense, hair shampoo, essential oil, medicinal.
Rubus rolfei	Emerald carpet raspberry	Sh	E	0.08	7-9	LMH	○ ⊕ ●	M	ANB		Edible fruit. Dye.
Rubus hispidus	Swamp dewberry	Sh	E	0.20	3-7	LMH	○ ⊕	M	ANB		Edible fruit. Dye.
Rubus illecebrosus	Strawberry-raspberry	Sh	D	0.60	4-8	LMH	○ ⊕	M	ANB	M	Edible fruit. Dye.
Rubus nepalensis	Nepalese raspberry	Sh	E	0.20	7-10	LMH	○ ⊕	M	ANB	M	Edible fruit. dye.
Rubus parvus		Sh	E	0.20	8-11	LMH	○ ⊕	M	ANB		Edible fruit, sap. Dye.
Rubus tricolor	Creeping bramble	Sh	E	0.30	6-9	LMH	○ ⊕ ●	DM	ANB	F	Edible fruit. dye.

Botanical Name	Common Name	Habit	Evergreen/ Deciduous	Height (meters)	US Hardiness Zone	Soil	Shade	Moisture	pH	Growth Rate	Notes
Salix repens	Creeping willow	Sh	D	1.50	4-8	LMH	○ ⊕	M	ANB		Soil stabilizer and soil reclamation.
Salvia officinalis	Sage	Sh	E	0.60	5-10	LM	○	DM	NB	M	Condiment, tea. Essential oil, insect repellent.
Sanicula europaea	Wood sanicle	P	D	0.60	5-9	LMH	○ ⊕	M	ANB		Edible leaves.
Santolina chamaecyparissus	Lavender-cotton	Sh	E	0.60	6-9	LM	○	DM	ANB	M	Condiment. Essential oil, insect repellent, dye.
Sasa veitchii	Kuma-zasa	B	E	1.50	7-10	LMH	⊕	M	ANB	F	
Saxifraga stolonifera	Strawberry saxifrage	P	D	0.40	6-10	LMH	⊕	M	ANB		Edible leaves and flowering stem.
Schizophragma hydrangeoides	Japanese Hydrangea Vine	Cl	D	6.00	5-8	LMH	○ ⊕	M	ANB	S	Edible leaves.
Sedum acre	Common stonecrop	P	E	0.12	4-9	LMH	○	DM	ANB	F	Edible leaves, condiment. Tolerates drought.
Sedum album	Small houseleek	P	E	0.10	6-8	LMH	○	DM	ANB	F	Edible leaves. Tolerates maritime exposure.
Sedum anacampseros	Loce restorer	P	E	0.10	5-9	LMH	○ ⊕	DM	AN		Edible leaves.
Sedum spathulifolium	Broadleaf Stonecrop	P	E	0.15	6-10	LMH	○	DM	ANB	M	Edible leaves.
Sedum spurium	Caucasian Stonecrop	P	E	0.15	6-9	LMH	○	DM	ANB	M	Edible leaves.
Shibataea kumasasa	Okame-zasa	B	E	0.75	6-9	LMH	⊕	M	ANB	F	

Botanical Name	Common Name	Habit	Evergreen/ Deciduous	Height (meters)	US Hardiness Zone	Soil	Shade	Moisture	pH	Growth Rate	Notes
Stephanandra incisa	Laceshrub	Sh	D	2.50	3-7	LMH	○ ⊕	M	ANB		Edible young leaves.
Symphytum tuberosum	Tuberous comfrey	P	D	0.60	4-8	LMH	○ ⊕	M	ANB	F	Coffee substitute.
Thymus praecox	Creeping thyme	Sh	E	0.05	5-8	LM	○	DM	ANB	M	Condiment. Essential oil, medicinal.
Thymus serpyllum	Wild thyme	Sh	E	0.10	5-9	LM	○	DM	ANB	M	Edible leaves, tea. Essential oil, insect repellent, fungicide, medicinal. Wind tolerant.
Thymus vulgaris	Common thyme	Sh	E	0.20	5-11	LM	○	DM	NB	M	As above.
Thymus x citriodorus	Lemon thyme	Sh	E	0.20	5-10	LM	○	DM	NB	M	As above.
Tiarella cordifolia	Foamflower	P	E	0.20	3-9	LMH	○ ⊕	M	ANB	S	Medicinal.
Tolmiea menziesii	Youth on age	P	D	0.60	7-9	LMH	⊕	DM	ANB	M	Edible young shoots.
Trachelospermum asiaticum	Japanese Star Jasmine	Cl	E	6.00	7-10	LMH	○ ⊕	M	ANB	S	
Trifolium repens	White clover	P	E	0.10	4-8	LMH	○	M	ANB	M	Edible leaves and flowers. Tea.
Vaccinium crassifolium	Creeping blueberry	Sh	E	0.10	6-9	LM	○ ⊕	M	A		Edible fruit.
Vaccinium nummularia	Nummularia-like blueberry	Sh	E	0.10	7-9	LM	○ ⊕	M	A		Edible fruit
Vaccinium praestans	Kamchatka bilberry	Sh	D	0.15	4-8	LM	⊕	M	A	S	Edible fruit.
Vaccinium vitis-idaea	Cowberry	Sh	E	0.35	3-8	LM	○ ⊕	M	A		Edible fruit, tea. Dye.

Botanical Name	Common Name	Habit	Evergreen/ Deciduous	Height (meters)	US Hardiness Zone	Soil	Shade	Moisture	pH	Growth Rate	Notes
Vinca minor	Lesser periwinkle	Sh	E	0.20	4-9	LMH	○ ⊕ ●	DM	ANB	M	Basket making. Medicinal. Drought tolerant.
Viola cornuta	Viola	P	E	0.30	6-11	LMH	○ ⊕	M	AN		Edible leaves, flowers, tea.
Viola labradorica	Labrador violet	P	E	0.10	3-8	LMH	○ ⊕ ●	M	AN		Edible leaves and flowers.
Viola obliqua	Marsh blue violet	P	E	0.15	3-8	LMH	○ ⊕	MWe	ANB		Edible flowers, leaves, tea.
Vitis coignetiae	Crimson glory vine	Cl	D	20.00	4-8	LMH	○ ⊕	DM	ANB	F	Edible leaves and stems. Dye.
Vitis davidii	Spiny vitis	Cl	D	20.00	6-9	LMH	○ ⊕	DM	ANB		Edible fruit and leaves. Dye.
Xanthorhiza simplicissima	Yellowroot	Sh	D	1.00	5-9	LMH	○ ⊕	MWe	A		Dye.

Table 3—plants suitable for hedging and windbreaks

Deciduous/evergreen
D = Deciduous
E = Evergreen

Height
S = Small (to 4ft for hedges, 20ft for trees)
M = Medium (to 6ft for hedges, 50ft for trees)
T = Tall (over 6ft for hedges, over 50ft for trees)

Rate of Growth
S = Slow
M = Medium
F = Fast

Wind Resistance
W = tolerates Windy sites
M = tolerates Maritime exposure

Botanical Name	Common Name	Deciduous/ Evergreen	Height	Growth Rate	US Hardiness Zone	Notes
Acer campestre	Field Maple	D	M-T	M	4-8	Most soils. Leaves preserve fruit. Medicinal
Ailanthus altissima	Tree of heaven	D	T	F	5-8	Tolerates most soils and conditions including pollution. Insect repellent, medicinal, soil stabilization, yellow dye. Leaves are said to be edible but slightly toxic
Alnus glutinosa	Alder	D	M-T	F	3-7	Prefers a moist soil. Ink ,dyes, tannin. Medicinal
Amelanchier canadensis	Juneberry	D	M-T	M-F	4-7	Moist well drained lime free soil, sunny position. Edible fruit
Arundo donax	Giant reed	Grass	M-T	F	6-10	Moist soil esp by water. Edible rhizome. Basketry, plant supports, erosion control. Medicinal
Atriplex canescens	Salt bush	E	M-T	M	6-9	Well drained soil in full sun. Resists fire. Edible leaves-salty. Edible seed
Atriplex halimus	Salt bush	E	M	M	7-10	As above
Aucuba japonica	Spotted laurel	E	M	M	7-10	Most soils sun or shade. Tolerates smoke. Leaves are said to be a famine food

Botanical Name	Common Name	Deciduous/Evergreen	Height	Growth Rate	US Hardiness Zone	Notes
Baccharis halimifolia	Bush groundsel	D	M-T	M	4-8	Any soil
Baccharis patagonica		E	M	S-M	7-10	Any soil
Bambusa multiplex	Bamboo	E	M-T	M	8-11	Rich moist soil, sheltered position. Edible shoots. Fibre for paper
Berberis amurensis	Barberry	D	M	M	5-9	Most soils
Berberis darwinii	Darwin's Barberry	E	M	S-M	7-9	Most soils. Edible fruit. Yellow dye. Medicinal
Berberis gagnepainii	Barberry	E	M	S-M	4-8	Most soils
Berberis soulieana	Barberry	E	M	M	5-9	Most soils
Berberis x stenophylla	Barberry	E	M	S-M	4-8	Most soils. Edible fruit
Berberis thunbergii	Japanese Barberry	D	M	M	4-8	Most soils. Edible fruit and leaves, not at all nice. Yellow dye. Medicinal
Berberis verruculosa	Barberry	E	M	S-M	4-8	Most soils
Berberis vulgaris	Common barberry	D	M	M	3-7	Most soils. Alternate host of black stem rust of cereals. Edible fruit, tea
Bupleurum fruticosum		E	M	M	6-9	Well drained soil
Bursaria spinosa	Christmas bush	E	M	S	7-10	Well drained moisture retentive soil. Used in suntan lotions to protect from U.V.light
Buxus sempervirens	Box	E	S-T	S	6-8	Most soils, sun or semi shade. Resists honey fungus. Medicinal
Caesalpinia decapetala	Mysore Thorn	D	M	S	7-10	Sheltered sunny position in well-drained soil. Tannin. Medicinal
Callistemon citrinus	Bottlebrush	E	M	M	8-11	Sheltered position. Tea. Tan & cinnamon dyes. C.sieberi & C.viridiflorus can also be used
Calluna vulgaris	Ling	E	S	S	4-7	Dry acid soil, sunny position. Basketry, thatching, brooms, yellow dye, tea. Medicinal. Fuel
Caragana arborescens	Siberian pea shrub	D	M-T	M	2-7	Sunny position, well-drained soil. Edible seed and young pods cooked. Erosion control, dye

CREATING YOUR PERMACULTURE HEAVEN 303

Botanical Name	Common Name	Deciduous/ Evergreen	Height	Growth Rate	US Hardiness Zone	Notes
Caragana brevispina	Long-stalked pea shrub	D	M	M	4-9	Sunny position, well-drained soil. Edible seed, cooked. Erosion control
Carpinus betulus	Hornbeam	D	M-T	M	5-7	Most soils. Yellow dye. Medicinal
Ceanothus thyrsiflorus	Blueblossom	E	M-T	F	7-9	Sunny position. Slow to establish. Soap from flowers, green dye
Cedrus libani atlantica	Atlas cedar	E	T	M	5-9	Most soils. Medicinal
Celastrus orbiculatus	Bittersweet	D	T	M	4-8	Deep rich loamy soil. Edible young leaves (cooked). Medicinal
Chaenomeles speciosa	Japanese quince	D	M	M	4-8	Most soils, sun or semi shade. Edible fruit. Medicinal
Chamaecyparis lawsoniana	Lawsons Cypress	E	T	F	5-7	Most soils. Resin
Choisya ternata	Mexican orange	D	M-T	M	6-9	Light soils
Colletia armata		D	M-T	S	7-10	Sandy well drained soil, sun or light shade
Cornus mas	Cornelian cherry	D	T	M	4-8	Most soils sunny position. Edible fruit. Oil, dye
Cornus sanguinea	Dogwood	D	M-T	M	4-8	Most soils. Oil, basketry, grey-blue dye. The fruit is said to be edible but you are welcome to it
Corokia x virgata		E	M-T	M	7-10	Most soils, well drained
Corylus avellana	Hazel nut	D	M-T	M	4-8	Most soils. Edible seed, oil. Basketry, wood polish, hurdles. Named varieties can also be used
Corylus maxima	Filbert	D	M-T	M	4-8	As above
Cotoneaster spp.	Cotoneaster	E	M-T	M	6-8	Most soils. Several species can be used inc;- C."Cornubia", C.divaricatus, C.franchetii, C.frigidus, C.glaucophyllus, C.lacteus, C.simonsii & C.wardii
Crataegus monogyna	Hawthorn	D	M-T	M-F	4-8	Most soils. Edible fruit, young shoots. Tea, coffee, medicinal. Also C.laevigata, C.mollis, C.sanguinea, & C.submollis

Botanical Name	Common Name	Deciduous/ Evergreen	Height	Growth Rate	US Hardiness Zone	Notes
Cupressocyparis leylandii	Leyland cypress	E	T	F	6-10	Most soils
Cupressus macrocarpa	Monterey cypress	E	T	F	7-10	Most soils but not shallow. Var."Lutea" is hardier
Drimys lanceolata	Mountain pepper	E	M-T	S-M	7-10	Semi shade. Dioecious. Fruit is a pepper substitute. Medicinal
Elaeagnus angustifolia	Oleaster	D	M-T	M-F	2-7	Prefers well drained soils. Resists drought and honey fungus. Fixes nitrogen. Edible fruit. Gum
Elaeagnus commutata	Silverberry	D	M	M	2-6	Light soil, sunny position. Resists drought and honey fungus. Fixes nitrogen. Edible fruit. Fibre
Elaeagnus x ebbingei	Ebbing's Silverberry	E	M-T	M-F	5-9	Best in not too rich a soil, tolerates shade. Fixes nitrogen. Edible fruit, produced in April/May
Elaeagnus glabra	Goat nipple	E	M-T	M	7-10	Most soils, fixes nitrogen. Very shade tolerant, will eventually climb into trees. Edible fruit
Elaeagnus macrophylla	Broad-leaved oleaster	E	M-T	M	6-9	As E.glabra
Elaeagnus multiflora	Goumi	D	M-T	M	5-9	As E.angustifolius. Tolerates air pollution
Elaeagnus pungens	Thorny olive	E	M-T	M	6-10	As E.glabra
Elaeagnus x reflexa		E	M-T	M	6-9	As E.glabra
Elaeagnus umbellata	Autumn olive	D	M-T	M	3-7	As E.angustifolius. Edible fruit and seed
Elaeagnus parvifolia	Autumn olive	D	M-T	M	3-7	As above but said to have a superior fruit
Eleutherococcus sieboldianus	Ukogi	D	M-T	M	4-8	Well drained humus rich soil, sunny position. Tolerates urban pollution. Edible leaves. Tea
Erica x darleyensis	Cape heath	E	S	M	6-9	Prefers a light peaty soil but tolerates chalk
Erica x veitchii	Heather	E	S-M	M	7-10	Light peaty soil

Botanical Name	Common Name	Deciduous/ Evergreen	Height	Growth Rate	US Hardiness Zone	Notes
Eucryphia lucida	Leatherwood	E	M-T	M	7-10	Most soils preferably lime-free. Dislikes cold winds. Sun or semi-shade. Medicinal
Eucryphia+x+nymansensis		E	M-T	M	6-9	Most soils preferably lime-free. Dislikes cold winds. Full sun only if soil is moist
Euonymus alatus	Spindle	D	M	M	4-8	Most soils if well drained. Medicinal, tea
Euonymus fortunei	Winter creeper	E	S	M	5-9	Most soils sun or shade
Euonymus japonicus	Japanese Spindle	E	M-T	M-F	6-9	Well drained soil, sun or shade. Hosts sugar beet fly. Rubber from root and stems
Fagus sylvatica	Common Beech	D	M-T	S-M	4-7	Most soils except heavy and wet. Edible young leaves, seed and oil. Medicinal
Forsythia x intermedia	Golden bell	D	M-T	M	5-8	Well drained preferably rich soil
Fuchsia magellanica	Fuchsia	D	M-T	M	5-7	Most soils. Often cut back to base in cold winters. Var."Ricartonii" is hardier. Dye, medicinal
Garrya elliptica	Coast silk tassel	E	M-T	M	7-10	Most well drained soils, not rich. Resents root disturbance. Grey to black dye. Medicinal
Gaultheria mucronata	Prickly heath	E	S	M	5-9	Light peaty acid soil, sunny position. Edible fruit
Genista hispanica	Spanish gorse	D	S	M	5-9	Sunny position in dry not rich soils. Resents root disturbance
Griselinia littoralis		E	M-T	M	6-9	Light rich soil, chalk tolerant. Edible fruit with a bitter flavour
Hakea sericea	Silky hakea	E	M	M	8-11	Moist well drained soil. Gum-similar to gum tragacanth
Hebe brachysiphon		E	S-M	M	6-9	Sunny, well drained
Hebe dieffenbachii		E	S-M	M	8-11	Sunny, well drained
Hebe x franciscana	Hebe	E	S-M	M	9-11	Light well drained soil on poor side, chalk tolerant

Botanical Name	Common Name	Deciduous/ Evergreen	Height	Growth Rate	US Hardiness Zone	Notes
Hebe speciosa	New Zealand hebe	E	S-M	M	6-9	As above
Helichrysum italicum	Curry plant	E	S	M	7-10	Well drained soil, sunny position. Leaves a flavouring
Hibiscus syriacus	Mallow	D	M	M	5-9	Most well drained soils, sunny position. Edible leaves, flowers, oil, tea, fibre, hair shampoo
Hippophae rhamnoides	Sea buckthorn	D	M-T	M	3-7	Most soils, sunny position. Fixes nitrogen. Edible fruit, very acid. Dyes. Medicinal
Hydrangea macrophylla	French Hydrangea	D	M	M	5-9	Well drained moist soil
Hydrangea serrata	Mountain Hydrangea	D	M	M	5-9	Well drained moist soil. Edible leaves, sweet in ssp. amagiana and thunbergii
Hymenanthera dentata	Tree violet	E	M-T	M	8-11	Well drained. Purple dye
Hypericum forrestii		D	S-M	M	4-8	Most soils
Hypericum patulum	Goldencup	E	S	M	6-7	Most soils
Hyssopus officinalis	Hyssop	E	S	M	5-10	Prefers light dry calcareous soil and a sunny position. Flavouring. Tea, medicinal
Ilex x altaclarensis	Holly	E	M-T	S	5-9	Most soils
Ilex aquifolium	English Holly	E	M-T	S	5-9	Most soils, not dry. Full sun to heavy shade. Medicinal
Laurus nobilis	Bay tree	E	M	S	8-10	Well drained fertile soil in full sun. Flavouring
Lavandula angustifolia	Lavender	E	S	M	5-8	Light dry warm soil, sunny position. Flavouring. Insect repellent, essential oil. Medicinal
Lavandula latifolia	Spike Lavender	E	S	M	6-9	As above
Leptospermum laevigatum	Coast tea tree	E	M-T	M	8-11	Most soils. Sand binder
Leptospermum scoparium	Tea tree	E	M-T	M	8-11	Most soils. Tea, green dye, insect repellent

Botanical Name	Common Name	Deciduous/ Evergreen	Height	Growth Rate	US Hardiness Zone	Notes
Ligustrum delavayanum	Delavay privet	E	M-T	M	6-9	Most soils, sun or shade
Ligustrum lucidum	Chinese privet	E	M-T	M	8-11	Most soils, tolerates shade and drought. Medicinal
Ligustrum ovalifolium	California Privet	E	M-T	M	4-8	Most soils, sun or shade
Ligustrum vulgare	Common Privet	D	M	M	4-7	Most soils, tolerates drought and shade. Basketry, ink dyes. Medicinal
Lonicera nitida	Boxleaf honeysuckle	E	S-M	M-F	6-9	Most soils
Luma apiculata	Temu	E	M-T	S-M	8-11	Most well drained soils. Edible fruit
Lycium barbarum	Goji	E	M	S-M	6-9	Sunny position, well-drained soil. Edible fruit and shoots. Medicinal
Lycium chinense	Chinese Boxthorn	E	M	S-M	5-9	Sunny position, well-drained soil. Edible fruit and shoots. Medicinal
Maclura pomifera	Osage orange	D	M-T	M	4-9	Most soils, sunny position, food preservative, insect repellent, tannin, yellow dye
Mahonia aquifolium	Oregon grape	E	S-M	M	4-8	Most soils, sun or heavy shade, edible fruit. Medicinal
Mahonia trifoliolata	Mexican barberry	E	M	S-M	6-9	Most soils preferably hot and dry, edible fruit, dyes, ink, tannin
Metasequoia glyptostroboides	Dawn redwood	D	T	M-F	5-8	Most soils but slow in dry sandy soils or shallow chalk. Soil stabilization
Miscanthus sacchariflorus	Amur silver grass	Grass	M-T	M-F	7-10	Most soils, not dry. Dies down in winter
Muehlenbeckia complexa	Maidenhair vine	D	M-T	M	7-10	Well drained soil in sun or semi-shade. Dioecious. Edible fruit
Myrtus communis	Myrtle	E	M-T	S-M	9-11	Most well drained soils. Flavouring. Essential oil, medicinal
Olearia spp.	Daisy bush	E	M-T	M	8-10	Most soils. Many species incl: O.aviceniifolia, O.x.haastii, O.illicifolia, O.macrodonta, O.solandri, O.traversii and O.virgata
Orixa japonica		D	M	M	5-9	Any soil, sun or shade. Edible young leaves-spicy

Botanical Name	Common Name	Deciduous/Evergreen	Height	Growth Rate	US Hardiness Zone	Notes
Osmanthus x burkwoodii	Burkwood osmanthus	E	M-T	S	5-9	Most soils esp chalk. Other species include: O.decorus, O.delavayi
Paliurus spina-christi	Christ's thorn	D	M-T	M	7-10	Most well drained soils, full sun
Perovskia atriplicifolia	Russian sage	D	S-M		5-10	Well-drained soils including chalk. Sunny position
Philadelphus coronarius	Mock orange	D	M	M	4-8	Most soils incl chalk. Sunny. Also: P.delavayi, P.pubescens, P.purpurascens & P.x.virginalis
Phillyrea latifolia	Green olive tree	E	M-T	S	6-9	Most soils. Edible fruit. Medicinal
Phormium tenax	New Zealand Flax	E	M-T	M	8-10	Most soils, preferably moist. Edible gum, nectar. Fibre, basketry, glue, dyes, tannin, coffee sub
Photinia davidiana	Christmas berry	E	M-T	F	7-9	Fertile well drained soil in a sheltered sunny position
Photinia x fraseri	Red tip photinia	E	T	M	7-9	As above. Succeeds on chalk
Pittosporum crassifolium	Karo	E	M-T	M	8-11	Most soils. Dark blue dye, saponins. Soil stabilization
Pittosporum ralphii	Ralph's desert willow	E	M-T	M	8-11	Most soils
Pittosporum tenuifolium	Kohuhu	E	M-T	M	7-10	Most soils. Edible gum. Foliage in flower arranging
Pittosporum tobira	Tobira	E	S	M	8-11	Well-drained soil, sun or semi-shade
Pittosporum undulatum	Cheesewood	E	M-T	M	9-11	Most soils
Pleioblastus hindsii	Bamboo	E	M-T	M	6-9	Most soils, not dry. Prefers shade. Edible young shoots. Plant supports
Pleioblastus simonii	Medake	E	M-T	M-F	5-9	As above
Poncirus trifoliata	Bitter orange	E	M-T	M	6-9	Most soils, sunny position. Edible fruit-a conserve. Medicinal
Potentilla fruticosa	Cinquefoil	D	S	M	2-6	Most soils, preferably light, well-drained. Tea, tinder, soil stabilization
Prinsepia utilis	Cherry prinsepia	D	M-T	M	6-9	Light soil, sunny position. Edible fruit, oil. Medicinal

Botanical Name	Common Name	Deciduous/ Evergreen	Height	Growth Rate	US Hardiness Zone	Notes
Prumnopitys andina	Plum-fruited yew	E	M-T	S-M	7-10	Most soils, sheltered position. Edible fruit-tastes like a grape. Edible seed
Prunus caroliniana	Laurel cherry	E	M	M	7-10	Well drained soil in sun or shade. Edible fruit
Prunus cerasifera	Cherry plum	D	T	M-F	5-8	Well drained, preferably with chalk. Sunny position. Edible fruit
Prunus cerasus	Sour cherry	D	T	M-F	3-7	Most well drained soils, preferably acid. Edible fruit, oil, gum, tea
Prunus x cistena	Purple-leaf sand cherry	D	M	M	4-8	Well drained soils including chalk, sunny position
Prunus incisa	Fuji cherry	D	M-T	M	5-7	Well drained soils including chalk, sunny position
Prunus insititia	Damson	D	T	M	5-9	Most soils incl chalky, sunny position. Edible fruit. Medicinal
Prunus laurocerasus	Cherry laurel	E	T	M	6-8	Most soils, dislikes shallow chalk. Shade tolerant. Edible fruit if fully ripe. Almond flavouring
Prunus lusitanica	Portuguese laurel	E	T	M	6-9	Most soils, sun or shade. Susceptible to silver leaf disease
Prunus spinosa	Sloe	D	M-T	M	4-8	Most well drained soils. Edible fruit, Ink, tea, dye, Medicinal
Pseudosasa japonica	Bamboo	E	T	M	5-9	Most soils. Edible young shoots. Plant support. Medicinal
Pyracantha spp.	Firethorn	E	M-T	M	4-10	Most soils. Susceptible to fire blight. Species include: P.coccinea, P.rogersiana, P."Wateri"
ilex	Holm oak	E	T	S-M	4-10	Most soils except cold poorly drained. Edible seed, oil. Slug repellent, tannin
Quercus petraea	Sessile oak	D	T	S-M	5-8	Most soils. Edible seed-coffee substitute. Slug repellent, tannin
Quercus robur	English Oak	D	T	S-M	4-8	As above
Rhamnus alaternus	Italian buckthorn	E	M	M	6-9	Well-drained soil

Botanical Name	Common Name	Deciduous/Evergreen	Height	Growth Rate	US Hardiness Zone	Notes
Rhamnus cathartica	Buckthorn	D	M-T	M	3-7	Most soils preferably calcareous. Primary host of Oat crown rust. Dyes. Medicinal
Rhamnus daburica	Dahurian buckthorn	D	M-T	M	4-8	Most soils. Tea, dye
Rhamnus frangula	Alder buckthorn	D	M-T	M	3-7	Moist fairly fertile soil, sun or part shade. Dyes. Medicinal
Rhaphiolepis umbellata	Japanese hawthorn	E	M-T	M	7-10	Most well-drained soils, sheltered position. Edible seed Brown dye
Rhododendron spp.	Rhododendron	E	M-T	M	7-9	Lime free peaty well drained soil. Species incl: R.lutescens, R.luteum, R.ponticum
Ribes sanguineum	Flowering currant	D	M-T	M	5-9	Most soils. Edible fruit
Rosa acicularis	Prickly rose	D	M	M	2-6	Most soils. Edible fruit, young shoots. Tea
Rosa glauca	Red leaved rose	D	M	M-F		Most soils
Rosa macrophylla		D	M-T	M-F	6-9	Most soils. Edible fruit
Rosa multiflora	Japanese rose	D	M-T	M-F	4-8	Most soils. Edible fruit, young shoots
Rosa nutkana	Nootka rose	D	M	M	4-8	Most soils. Edible fruit, petals, young shoots. Tea
Rosa rubiginosa	Eglantine	D	M-T	M-F	4-8	Most soils. Edible fruit, petals, young shoots. Tea. Medicinal
Rosa rugosa	Ramanas rose	D	M	M-F	3-9	Most soils. Edible fruit, petals. Tea. Medicinal
Rosa virginiana	Virginian rose	D	M	M	3-7	Most soils. Edible fruit, young shoots. Stabilizing sand dunes
Salix alba	White willow	D	T	F	2-8	Prefers a stiff moist soil. Edible leaves. Basketry, medicinal
Salix "Bowles hybrid"	Bowles willow	D	T	F	8-11	Prefers a stiff moist soil. Basketry
Salix caprea	Goat willow	D	T	F	4-9	Most soils if moist. Basketry, leather substitute
Salix lanata	Wooly willow	D	S-M	M	4-7	Most soils
Salix purpurea	Purple osier	D	T	F	4-8	Most soils. Basketry. Medicinal

Botanical Name	Common Name	Deciduous/ Evergreen	Height	Growth Rate	US Hardiness Zone	Notes
Sambucus nigra	Elder	D	M-T	F	5-7	Most soils. Edible fruit, flowers. Insect repellent. Tea Medicinal
Santolina neopolitanum	Holy flax	E	S	M	6-9	Sunny position, poor soil
Santolina rosmarinifolia		E	S	M	6-10	As above
Sasa palmata	Broadleaf Bamboo	E	M	M	7-8	Moist well drained loam. Plant supports
Semiarundinaria fastuosa	Narihira Bamboo	E	T	M	6-9	Moist well-drained loam, sheltered position. Edible young shoots. Plant supports
Senecio spp.		E	M-T	S-M	9-11	Most soils. Species incl: S.greyi, S.laxifolius, S.monroi, S.reinoldii, S."Sunshine"
Shepherdia argentea	Buffalo berry	D	M-T	M	3-9	Most soils, drought resistant. Edible fruit. Red dye
Sinarundinaria nitida	Fountain bamboo	E	T	M	4-8	Moist well drained loam, best in semi shade. Plant supports
Smilax aspera	Sarsaparilla	E	M-T	M	8-11	Most soils. Edible young shoots. Root used in soft drinks. Dy. Medicinal
Sorbus intermedia	Swedish whitebeam	D	M-T	M	4-8	Most well-drained soils, preferably moist. Sunny position. Edible fruit
Spartium junceum	Spanish broom	D	M-T	F	8-9	Most soils, preferably lime free. Basketry, fibre, dye, essential oil. Medicinal
Spiraea spp.	Bridal wreath	D	M	M	4-8	Most moist soils, sunny position. Species include: S.x.arguta, S.japonica and S.thunbergii
Symphoricarpos albus laevigatus	Snowberry	D	M-T	M-F	3-7	Most soils and situations. Invasive. Edible fruit
Syringa vulgaris	Lilac	D	M-T	M-F	3-7	Most soils, warm sunny position. Essential oil, dye. Medicinal
Tamarix gallica	Tamarisk	D	T	M	4-8	Most soils. Edible manna. Medicinal. Also T.africana, T.parviflora and T.ramosissima

Botanical Name	Common Name	Deciduous/ Evergreen	Height	Growth Rate	US Hardiness Zone	Notes
Taxus baccata	Yew	E	M-T	S-M	5-7	Most soils, sun or shade. Edible fruit-sweet. All other parts of plant are poisonous. Insecticide, incense. Medicinal. Other species can also be used
Thamnocalamus spathaceus	Umbrella bamboo	E	T	M	5-9	Moist well drained loam. Edible young shoots. Plant supports
Thuja occidentalis	White cedar	E	T	S-M	3-7	Moist soil. Best in west of Britain. Tea, insect repellent. Medicinal
Thuja orientalis	Biota	E	M-T	S-M	5-9	Prefers a dry well drained soil. Often does not do well. Edible seed. Dye. Medicinal
Thuja plicata	Red cedar	E	M-T	M-F	5-8	Most soils except light, sandy. Edible inner bark. Fibre, basketry
Tsuga heterophylla	Western Hemlock	E	T	M-F	6-7	Best in high rainfall areas in a deep well drained soil. Edible inner bark. Tea, tannin, resin. Medicinal. Other species to use include: T.canadensis, T.caroliniana, T.chinensis and T.mertensiana
Ugni molinae	Ugni	E	M	S-M	7-11	Most well-drained soils. Edible fruit. Tea
Ulex europaeus	Gorse	E	M	F	5-9	Prefers a poor dry soil, sunny position. Soil stabilization, tea, dye, kindling
Ulex parviflorus		E	M	M-F	6-9	Prefers a poor dry soil, sunny position. Soil stabilization, tea, dye, kindling
Viburnum x juddii	Judd viburnum	D	M	M	4-8	Dislikes poor dry soils
Viburnum opulus	Guelder rose	D	M-T	M-F	3-8	Prefers moist soils. Edible fruit-best cooked. Red dye, medicinal. Ssp.americanum is said to have superior fruit
Viburnum prunifolium	Black haw	D	M-T	M	3-9	Tolerates poor dry soils. Edible fruit
Viburnum tinus	Laurustinus	E	M	F	8-9	Most soils
Vitex negundo	Chinese chastetree	D	M-T	M	6-9	Light soil, sunny position. Tea, insect repellent, basketry. Medicinal
Weigela florida	Weigela	D	M	M	4-8	Rich soil, sun or part shade

Table 4—plants for the food forest

Plants for the food forest				
Botanical Name	Common Name	Edible Rating	Medicinal Rating	USDA Hardiness Zone
Acacia angustissima	Prairie acacia, Timbre, Fernleaf Acacia	2	3	7-10
Acer saccharinum	Silver Maple, River Maple, Soft Maple	3	1	3-9
Acer saccharum	Sugar Maple, Florida Maple, Hard Maple, Rock Maple	4	2	4-8
Acer saccharum nigrum	Black Maple	4	1	4-6
Achillea millefolium	Yarrow, Boreal yarrow, California yarrow, Giant yarrow, Coast yarrow, Western yarrow, Pacific yarrow	3	4	4-8
Achillea ptarmica	Sneeze-Wort, Sneezeweed	2	1	3-9
Actinidia arguta	Tara Vine	5	0	4-8
Actinidia deliciosa	Kiwi Fruit	5	1	6-9
Actinidia kolomikta	Kiwi	4	0	3-8
Actinidia purpurea	Purple hardy kiwi	4	0	5-9
Adiantum pedatum	Northern Maidenhair, American Maidenhair Fern	0	2	4-9
Agastache foeniculum	Anise Hyssop, Blue giant hyssop	5	1	4-9
Alcea rosea	Hollyhock	3	2	5-9
Allium ampeloprasum	Wild Leek, Broadleaf wild leek	5	3	5-9
Allium canadense	Canadian Garlic, Meadow garlic, Fraser meadow garlic, Hyacinth meadow garlic	4	2	4-8
Allium cepa aggregatum	Potato Onion	4	3	4-8

Plants for the food forest				
Botanical Name	Common Name	Edible Rating	Medicinal Rating	USDA Hardiness Zone
Allium cepa proliferum	Tree Onion, Walking Onion	5	3	4-8
Allium cernuum	Nodding Onion, New Mexican nodding onion	5	2	5-9
Allium fistulosum	Welsh Onion	5	2	5-9
Allium schoenoprasum	Chives, Wild chives, Flowering Onion	5	2	5-11
Allium tricoccum	Wood Leek, Ramp	4	2	5-9
Allium tuberosum	Garlic Chives, Chinese chives, Oriental Chives,	5	2	4-8
Alnus cordata	Italian Alder	0	0	5-9
Alnus incana	Grey Alder, Speckled alder, Thinleaf alder, White Alder	0	0	2-6
Alnus rugosa	Speckled Alder	0	2	2-6
Alnus serrulata	Smooth Alder, Hazel alder	0	2	3-9
Althaea officinalis	Marsh Mallow, Common marshmallow	5	5	3-7
Amelanchier alnifolia	Saskatoon, Saskatoon serviceberry, Serviceberry	5	2	4-6
Amelanchier arborea	Downy Serviceberry, Alabama serviceberry, Juneberry, Common Serviceberry, Downy Serviceberry	3	1	5-8
Amelanchier bartramiana	Oblongfruit serviceberry	3	0	4-8
Amelanchier canadensis	Juneberry, Canadian serviceberry, Serviceberry Downy, Shadblow, Shadbush, Serviceberry	4	1	4-7
Amelanchier lamarckii	Apple Serviceberry	5	0	3

Plants for the food forest				
Botanical Name	Common Name	Edible Rating	Medicinal Rating	USDA Hardiness Zone
Amelanchier obovalis	Southern Juneberry, Coastal serviceberry	3	0	5-9
Amelanchier stolonifera	Quebec Berry, Running serviceberry	5	1	4-8
Amorpha canescens	Lead Plant	2	2	2-9
Amorpha fruticosa	False Indigo, False indigo bush	1	2	4-8
Amorpha nana	Dwarf Indigobush, Dwarf false indigo, Dwarf Indigo	0	1	4-8
Amorphophallus konjac	Devil's Tongue, Devil's Tongue, Snake Plant, Konjac, Konnyaku Potato, Voodoo Lily	4	2	6-11
Amphicarpaea bracteata	Hog Peanut, American hogpeanut	5	1	4-9
Andropogon gerardii	Big Bluestem	0	1	4-8
Antennaria dioica	Catsfoot, Stoloniferous pussytoes	0	2	4-8
Apios americana	Ground Nut	5	1	3-7
Apios fortunei	Hodo, Hodoimo	5	1	4-9
Apios priceana	Traveler's delight	5	0	6-9
Aquilegia canadensis	Wild Columbine, Red columbine, Meeting Houses, Common Columbine	2	2	4-10
Arabis caucasica	Rock Cress, Wall Rockcress	2	0	4-9
Aralia cordata	Udo	4	2	4-9
Aralia elata	Japanese Angelica Tree, Angelica Tree	3	2	4-9
Aralia hispida	Bristly Sarsaparilla	2	1	3-7
Aralia nudicaulis	Wild Sarsaparilla	4	3	4-8

Plants for the food forest				
Botanical Name	Common Name	Edible Rating	Medicinal Rating	USDA Hardiness Zone
Aralia racemosa	American Spikenard	3	3	4-8
Araucaria araucana	Monkey Puzzle Tree	5	1	7-11
Arbutus unedo	Strawberry Tree	4	2	7-11
Arctostaphylos uva-ursi	Bearberry	3	4	4-8
Armoracia rusticana	Horseradish, Red Cole	3	3	4-9
Cacalia atriplicifolia	Pale indian plantain	2	1	4-6
Aronia arbutifolia	Red Chokeberry	2	0	4-9
Aronia melanocarpa	Black Chokeberry, Black Berried Aronia	3	1	3-8
Artemisia dracunculus	Tarragon, French Tarragon	4	2	5-8
Artemisia frigida	Fringed Wormwood, Prairie sagewort	2	2	3-10
Artemisia stelleriana	Beach Wormwood, Oldwoman, Dusty Miller	1	0	3-9
Arundinaria gigantea	Canebrake bamboo, Cane Reed, Giant cane	3	1	5-9
Asarum canadense	Snake Root, Canadian wildginger, Canada Wild Ginger, Wild Ginger	3	3	3-9
Asarum shuttleworthii	Asarabacca, Mottled Wild Ginger	2	0	5-9
Asclepias incarnata	Swamp Milkweed, Swamp Butterfly Weed, Marsh Milkweed	3	2	3-8
Asclepias syriaca	Common Milkweed, Silkweed, Milkweed	3	2	3-8
Asimina triloba	Papaw	4	2	5-8
Asparagus officinalis	Asparagus, Garden asparagus	4	3	2-9
Asphodeline lutea	King's Spear, Yellow Asphodel, Jacob's Rod	4	0	6-9

Plants for the food forest				
Botanical Name	Common Name	Edible Rating	Medicinal Rating	USDA Hardiness Zone
Aster novae-angliae	New England Aster	0	2	4-9
Astragalus canadensis	Canadian Milkvetch, Shorttooth Canadian milkvetch, Morton's Canadian milkvetch	3	2	7-10
Astragalus crassicarpus	Ground Plum, Groundplum milkvetch	4	1	6-9
Astragalus glycyphyllos	Milk Vetch, Licorice milkvetch	1	0	3-7
Astragalus membranaceus	Huang Qi	0	5	5-9
Athyrium filix-femina	Lady Fern, Common ladyfern, Subarctic ladyfern, Asplenium ladyfern, Southern Lady Fern, Tatting Fer	1	2	3-8
Atriplex halimus	Sea Orach, Saltbush	5	1	7-10
Balsamorhiza sagittata	Oregon Sunflower, Arrowleaf balsamroot	4	2	4-8
Baptisia australis	Wild Indigo, Blue wild indigo, Blue False Indigo	0	2	3-9
Baptisia tinctoria	Wild Indigo, Horseflyweed	1	3	4-8
Bellis perennis	Daisy, Lawndaisy, English Daisy	2	3	4-8
Berberis canadensis	Allegheny Barberry, American barberry	3	2	4-8
Beta vulgaris maritima	Sea Beet	2	2	4-8
Betula alleghaniensis	Yellow Birch, Swamp Birch	3	2	3-7
Betula lenta	Cherry Birch, Sweet birch, Black Birch, Cherry Birch	3	3	3-7
Borago officinalis	Borage, Common borage, Cool-tankard, Tailwort	3	3	6-9

Plants for the food forest				
Botanical Name	Common Name	Edible Rating	Medicinal Rating	USDA Hardiness Zone
Brassica oleracea	Wild Cabbage, Broccoli, Tronchuda cabbage, Brussels sprouts, Kohlrabi, Sprouting broccoli	4	2	6-9
Bunias orientalis	Turkish Rocket, Turkish warty cabbage	4	0	6-9
Bunium bulbocastanum	Pig Nut, Earth-nut	4	1	4-8
Calamintha nepeta	Lesser Calamint	3	2	5-9
Callirhoe involucrata	Poppy Mallow, Purple poppymallow, Winecup, Finger Poppy Mallow	3	2	4-8
Calycanthus floridus	Carolina Allspice, Eastern sweetshrub, Strawberry Bush, Sweetshrub, Carolina Allspice	3	2	5-10
Camassia leichtlinii	Wild Hyacinth, Large camas, Suksdorf's large camas	4	0	3-7
Camassia quamash	Quamash, Small camas, Utah small camas, Walpole's small camas	5	1	3-7
Campanula carpatica	Tussock Bellflower, Carpathian Bellflower, Carpathian Harebell	3	0	3-8
Campanula glomerata	Clustered Bellflower, Dane's blood, Clustered Bellflower	4	0	4-9
Campanula portenschlagiana	Adria Bellflower	3	0	3-7
Caragana arborescens	Siberian Pea Tree, Siberian peashrub	5	1	2-7
Caragana pygmaea	Pygmy Peashrub	1	0	3-7

Plants for the food forest				
Botanical Name	Common Name	Edible Rating	Medicinal Rating	USDA Hardiness Zone
Cardamine bulbosa	Bulbous Bittercress	2	0	4-8
Dentaria diphylla	Crinkleroot	4	2	3-8
Cardamine pratensis	Cuckoo Flower	3	2	4-8
Carya illinoinensis	Pecan	4	1	5-9
Carya laciniosa	Shellbark Hickory	3	1	5-9
Carya ovata	Shagbark Hickory	3	1	4-8
Carya hybrids	Hybrid and neohybrid hickories	4	3	4-11
Castanea dentata	American Sweet Chestnut	3	1	4-8
Castanea mollissima	Chinese Chestnut	3	2	4-8
Castanea pumila	Chinquapin, Ozark chinkapin	4	1	4-8
Ceanothus americanus	New Jersey Tea, Wild Snowball	3	3	4-9
Ceanothus prostratus	Squaw Carpet, Prostrate ceanothus	0	0	6-9
Celtis laevigata	Sugarberry, Netleaf hackberry, Texan sugarberry, Sugar Hackberry	2	1	5-10
Celtis occidentalis	Hackberry, Common hackberry	3	1	3-9
Centranthus ruber	Red Valerian, Fox's Brush, Jupiter's Beard	2	1	5-8
Cephalotaxus harringtonia	Japanese Plum Yew	5	0	6-9
Cercis canadensis	Redbud, Eastern redbud, Mexican redbud, Texas redbud	3	2	4-9
Cercocarpus montanus	Mountain Mahogany, Alderleaf mountain mahogany, Silver mountain mahogany, Island mountain mahogany,	0	1	6-7

Plants for the food forest				
Botanical Name	Common Name	Edible Rating	Medicinal Rating	USDA Hardiness Zone
Chaenomeles speciosa	Japanese Quince, Flowering quince	3	2	4-8
Chamaemelum nobile	Chamomile, Roman chamomile	2	5	4-8
Chasmanthium latifolium	Indian Woodoats, Wild Oats Grass, North American Wild Oats, Northern Sea Oats, Spanglegrass River Oa	1	0	4-10
Chenopodium bonus-henricus	Good King Henry	4	2	4-8
Chimaphila maculata	Spotted Wintergreen, Striped prince's pine, Pipsissewa	1	3	6-7
Chimaphila umbellata	Pipsissewa	2	3	4-8
Chrysosplenium americanum	Water Mat, American golden saxifrage	1	0	3-7
Cichorium intybus	Chicory, Radicchio, Succory, Witloof	4	3	3-7
Claytonia caroliniana	Broad-Leaved Spring Beauty, Carolina springbeauty	3	0	5-9
Claytonia megarhiza	Alpine Spring Beauty	3	0	4-8
Claytonia virginica	Spring Beauty, Virginia springbeauty, Hammond's claytonia, Yellow Virginia springbeauty	3	1	5-7
Clintonia borealis	Bluebeard	2	1	3-7
Colocasia esculenta	Taro, Elephant Ears Taro, Dasheen, Eddo	4	2	9-11
Colutea arborescens	Bladder Senna	0	2	4-8
Commelina erecta	Slender Day-Flower, Whitemouth dayflower	2	0	8-11

Plants for the food forest				
Botanical Name	Common Name	Edible Rating	Medicinal Rating	USDA Hardiness Zone
Comptonia peregrina	Sweet Fern	3	3	3-6
Conopodium majus	Pignut	3	0	6-9
Coptis trifolia	Goldthread, Threeleaf goldthread	2	2	2-7
Coreopsis auriculata	Tickseed, Lobed tickseed, Mouse-eared Coreopsis, Eared Coreopsis	0	0	4-8
Cornus canadensis	Creeping Dogwood, Bunchberry dogwood, Bunchberry	4	2	2-7
Cornus florida	Flowering Dogwood	2	2	5-9
Cornus kousa	Japanese Dogwood, Kousa dogwood, Chinese Dogwood,	5	0	5-8
Cornus mas	Cornelian Cherry, Cornelian Cherry Dogwood	4	2	4-8
Cornus sericea	Red Osier Dogwood, Western dogwood	2	2	2-7
Corylus americana	American Hazel	3	1	4-8
Corylus avellana	Common Hazel, Common filbert, European Filbert, Harry Lauder's Walking Stick, Corkscrew Hazel, Hazel	5	2	4-8
Corylus chinensis	Chinese Hazel	2	0	5-9
Corylus colurna	Turkish Hazel, Chinese hazelnut, Turkish Filbert, Turkish Hazel	3	1	4-7
Corylus cornuta	Beaked Hazel, California hazelnut, Turkish Filbert, Turkish Hazel	3	1	4-7
Crambe maritima	Sea Kale	4	0	4-8

Plants for the food forest				
Botanical Name	Common Name	Edible Rating	Medicinal Rating	USDA Hardiness Zone
Crataegus aestivalis	Eastern Mayhaw, May hawthorn, Mayhaw, Apple Hawthorn	3	2	6-11
Crataegus mollis	Red Haw, Downy hawthorn	4	2	4-8
Crataegus pinnatifida	Chinese Haw	3	3	5-9
Crataegus punctata	Dotted Hawthorn,	3	2	4-8
Crocus sativus	Saffron	3	3	5-9
Cryptotaenia canadensis	Honewort, Canadian honewort	3	0	4-8
Cryptotaenia japonica	Mitsuba, Japanese honewort	4	1	4-8
Cudrania tricuspidata	Silkworm Thorn, Storehousebush	2	2	6-9
Cunila origanoides	Stone Mint, Common dittany	2	2	5-9
Cydonia oblonga	Quince	4	2	5-9
Darmera peltata	Umbrella Plant, Indian rhubarb, Indian Rubarb, Indian Rubarb	2	0	5-8
Decaisnea fargesii	Blue Sausage Fruit	3	0	4-8
Dioscorea batatas	Chinese Yam	5	5	4-11
Dioscorea japonica	Glutinous Yam, Japanese yam	4	2	7-12
Diospyros kaki	Persimmon, Japanese persimmon	4	3	7-10
Diospyros virginiana	American Persimmon, Common persimmon, Persimmon	5	1	4-8
Diplotaxis muralis	Wall Rocket, Annual wallrocket	3	0	6-9
Diplotaxis tenuifolia	Perennial Wall Rocket	4	0	5-9

Plants for the food forest				
Botanical Name	Common Name	Edible Rating	Medicinal Rating	USDA Hardiness Zone
Dryas octopetala	Mountain Avens, Eightpetal mountain-avens, Alaskan mountain-avens, Hooker's mountain-avens, Kamtsch	1	1	3-6
Dryopteris marginalis	Marginal Woodfern, Leather Wood Fern	0	4	3-8
Echinacea purpurea	Echinacea, Eastern purple coneflower, Hedge Coneflower, Black Sampson, Purple Coneflower	1	5	3-10
Elaeagnus commutata	Silverberry	3	2	2-6
Elaeagnus multiflora	Goumi, Cherry silverberry	5	2	5-9
Epigaea repens	Mayflower, Trailing arbutus, Ground Laurel	2	2	3-8
Epilobium angustifolium	Willow Herb	3	2	3-7
Epilobium latifolium	River Beauty	3	2	4-8
Equisetum sylvaticum	Wood Horsetail, Woodland horsetail	1	2	3-11
Erigenia bulbosa	Harbinger Of Spring	2	1	4-7
Eryngium maritimum	Sea Holly, Seaside eryngo	3	3	4-8
Erythronium americanum	Trout Lily, Dogtooth violet	4	1	3-7
Eucalyptus gunnii	Cider Gum	3	3	7-10
Fagus grandifolia	American Beech	2	2	4-8
Fagus sylvatica	Beech, European beech, Common Beech	4	2	4-7
Foeniculum vulgare	Fennel, Sweet fennel	5	3	3-10
Fragaria chiloensis	Beach Strawberry, Pacific beach strawberry, Sandwich beach strawberry	3	1	4-10

Plants for the food forest				
Botanical Name	Common Name	Edible Rating	Medicinal Rating	USDA Hardiness Zone
Fragaria moschata	Hautbois Strawberry	3	0	5-9
Fragaria vesca	Wild Strawberry, Woodland strawberry, California strawberry	3	3	5-8
Fragaria virginiana	Scarlet Strawberry, Virginia strawberry	3	2	3-7
Fragaria x ananassa	Strawberry	5	0	4-8
Fritillaria camschatcensis	Kamchatka Lily, Kamchatka fritillary	4	0	4-8
Galax urceolata	Beetleweed, Wandflower	0	1	5-8
Galium odoratum	Sweet Woodruff, Sweetscented bedstraw, Bedstraw	3	3	5-9
Gaultheria hispidula	Creeping Snowberry	4	1	5-9
Gaultheria procumbens	Checkerberry, Eastern teaberry, Teaberry, Creeping Wintergreen	4	3	3-6
Gaultheria shallon	Shallon, Salal	5	2	6-9
Gaylussacia baccata	Black Huckleberry	4	1	5-9
Gaylussacia brachycera	Box Huckleberry	2	0	5-9
Genista tinctoria	Dyer's Greenweed, Common Woadwaxen, Broom	1	2	4-7
Geranium maculatum	Spotted Cranesbill, Spotted geranium, Crowfoot, Wild Geranium, Cranesbill	0	3	3-10
Geum rivale	Water Avens, Purple avens	3	2	3-7
Geum urbanum	Wood Avens, Bennet's Root— Old man's whiskers, Herb bennet	3	3	5-9
Ginkgo biloba	Maidenhair Tree, Ginkgo	5	5	3-8

Plants for the food forest				
Botanical Name	Common Name	Edible Rating	Medicinal Rating	USDA Hardiness Zone
Gleditsia triacanthos	Honey Locust	3	2	4-8
Glycyrrhiza glabra	Liquorice, Cultivated licorice	4	4	7-10
Glycyrrhiza lepidota	American Liquorice	4	3	3-8
Glycyrrhiza uralensis	Gan Cao	3	4	5-9
Hedysarum boreale	Sweet Vetch, Utah sweetvetch, Northern sweetvetch	4	0	3-7
Helianthus giganteus	Giant Sunflower	3	0	4-8
Helianthus maximilianii	Maximillian Sunflower, Maximillian Daisy	3	0	5-10
Helianthus tuberosus	Jerusalem Artichoke	4	1	4-8
Helianthus laetiflorus	Showy Sunflower, Cheerful sunflower	2	0	4-8
Hemerocallis fulva	Common Day Lily, Orange daylily, Tawny Daylily, Double Daylily	5	2	3-10
Hemerocallis lilioasphodelus	Yellow Day Lily	4	2	4-8
Heracleum sphondylium	Cow Parsnip, Eltrot	3	2	4-8
Heuchera americana	Rock Geranium, American alumroot, Alumroot, Coral Bells, Rock Geranium	0	3	4-9
Hibiscus syriacus	Rose Of Sharon, Althaea, Shrub Althea, Hardy Hibiscus	4	2	5-9
Hieracium venosum	Rattlesnake Weed	0	1	3-7
Hippophae rhamnoides	Sea Buckthorn, Seaberry	5	5	3-7
Houttuynia cordata	Tsi, Chameleon, Rainbow Plant, Chameleon Plant	4	3	5-10
Hovenia dulcis	Japanese Raisin Tree	3	2	5-9

Plants for the food forest				
Botanical Name	Common Name	Edible Rating	Medicinal Rating	USDA Hardiness Zone
Humulus lupulus	Hop, Common hop, European Hop,	4	5	5-7
Hydrophyllum virginianum	Virginia Waterleaf, Eastern waterleaf	3	1	4-8
Sedum telephium	Orpine	1	2	4-8
Ilex glabra	Inkberry	1	0	3-7
Illicium floridanum	Aniseed Tree, Florida anisetree, Purple Anise, Star Anise, Florida anise	1	0	6-10
Indigofera decora	Chinese indigo	2	1	5-7
Ipomoea leptophylla	Bush Moon Flower	3	2	8-11
Ipomoea pandurata	Wild Potato Vine, Man of the earth	3	2	6-9
Iris cristata	Crested Iris, Dwarf crested iris	1	1	5-9
Jeffersonia diphylla	Twinleaf, Rheumatism Root	0	2	5-8
Juglans ailanthifolia	Japanese Walnut	3	1	4-8
Juglans cinerea	Butternut—White Walnut, Butternut	3	3	3-7
Juglans nigra	Black Walnut	3	3	4-9
Juglans regia	Walnut, English walnut, Persian Walnut,	4	3	7-9
Juglans x bisbyi	Buartnut	3	0	4-8
Lactuca perennis	Perennial Lettuce	3	2	5-9
Laportea canadensis	Canadian Wood Nettle	3	1	4-8
Lathyrus japonicus maritimus	Beach Pea	2	0	3-7
Lathyrus latifolius	Perennial Sweet Pea, Perennial pea	1	0	5-9
Lathyrus linifolius montanus	Bitter Vetch	2	0	5-9
Lathyrus tuberosus	Earthnut Pea, Tuberous sweetpea	5	0	5-9

Plants for the food forest				
Botanical Name	Common Name	Edible Rating	Medicinal Rating	USDA Hardiness Zone
Lespedeza bicolor	Lespedeza, Shrub lespedeza	3	0	4-8
Levisticum officinale	Lovage, Garden lovage	3	3	5-9
Ligusticum scoticum	Scottish Lovage, Scottish licorice-root, Hulten's licorice-root	3	2	4-8
Lilium brownii	Hong Kong Lily	3	2	4-8
Lilium canadense	Meadow Lily, Canada lily	3	1	4-8
Lilium lancifolium	Tiger Lily, Devil Lily	4	2	4-8
Lilium longiflorum	White Trumpet Lily, Easter lily, Trumpet Lily	3	2	7-9
Lilium philadelphicum	Wood Lily	3	1	4-8
Lilium superbum	Swamp Lily, Turk's-cap lily, American Turk's Cap Lily	3	0	4-9
Lindera benzoin	Spice Bush, Northern spicebush, Bush Northern Spice	3	3	4-9
Linnaea borealis	Twinflower, Longtube twinflower	1	1	2-6
Lobelia cardinalis	Cardinal Flower	0	3	3-9
Lomatium cous	Biscuitroot, Cous biscuitroot	4	0	5-12
Lomatium dissectum	Fernleaf Biscuitroot, Carrotleaf biscuitroot	4	2	6-10
Lomatium macrocarpum	Bigseed Biscuitroot	4	2	5-10
Lomatium nudicaule	Pestle Parsnip, Barestem biscuitroot	4	2	6-8
Lonicera sempervirens	Trumpet Honeysuckle, Coral Honeysuckle	0	1	4-9
Lonicera villosa	Mountain fly honeysuckle, Fuller's honeysuckle	3	0	3-9
Lupinus perennis	Sundial Lupine	3	1	4-8

Plants for the food forest				
Botanical Name	Common Name	Edible Rating	Medicinal Rating	USDA Hardiness Zone
Lycopodium clavatum	Common Club Moss, Running clubmoss	0	3	10-12
Lycopus uniflorus	Bugleweed, Northern bugleweed	3	1	4-8
Maackia amurensis	Chinese Yellow Wood, Amur maackia	1	0	4-7
Magnolia virginiana	Laurel Magnolia, Sweetbay	1	3	4-8
Mahonia aquifolium	Oregon Grape, Hollyleaved barberry, Oregon Holly Grape, Oregon Holly	3	3	4-8
Mahonia repens	Creeping Oregon Grape, Creeping barberry, Grape Oregon	3	3	4-8
Maianthemum canadense	Canada Beadruby, Canada mayflower	1	1	3-7
Malus baccata	Chinese Crab, Siberian crab apple	2	1	2-7
Malus ioensis	Prairie Crab, Prairie crab apple, Texas crab apple, Prairie Crabapple	2	0	3-8
Malva alcea	Vervain mallow, Hollyhock Mallow	5	1	4-8
Malva moschata	Musk Mallow	5	2	3-10
Matteuccia struthiopteris	Ostrich Fern	2	1	2-7
Medeola virginiana	Indian Cucumber Root	4	1	3-7
Medicago sativa	Alfalfa, Yellow alfalfa	4	3	4-8
Melissa officinalis	Lemon Balm, Common balm, Bee Balm, Sweet Balm, Lemon Balm	3	5	4-8
Mentha arvensis	Corn Mint, Wild mint	3	2	4-8
Mentha requienii	Corsican Mint, Mint	3	2	5-9
Mentha spicata	Spearmint	4	3	3-9

Plants for the food forest				
Botanical Name	Common Name	Edible Rating	Medicinal Rating	USDA Hardiness Zone
Mentha suaveolens	Round-Leaved Mint, Apple mint, Pineapple Mint	2	2	5-10
Mentha x piperita officinalis	White Peppermint	3	5	3-7
Mentha x villosa alopecuroides	Apple Mint, Bowles' Mint	4	2	4-8
Mertensia maritima	Oyster Plant	3	0	3-7
Mespilus germanica	Medlar	4	1	5-8
Mitchella repens	Partridge Berry	3	3	4-9
Mitella diphylla	Mitrewort, Twoleaf miterwort	0	1	3-7
Monarda didyma	Bergamot, Scarlet beebalm, Horsemint, Oswego Tea, Bee Balm	3	2	4-10
Monarda fistulosa	Wild Bergamot, Mintleaf bergamot, Wild Bee-Balm, Lupine	3	2	4-10
Claytonia perfoliata	Miner's Lettuce	4	1	6-10
Claytonia sibirica	Pink Purslane, Siberian spring beauty	4	1	3-7
Morus alba	White Mulberry, Common Mulberry,	4	3	4-9
Morus nigra	Black Mulberry	5	3	5-9
Morus rubra	Red Mulberry, Common Mulberry, White Mulberry	3	2	4-9
Morus species	Mulberry	4	0	4-8
Myrica cerifera	Wax Myrtle—Bayberry Wild Cinnamon, Southern Bayberry, Wax Myrtle, Southern Wax Myrtle	3	3	7-11
Myrica gale	Bog Myrtle, Sweetgale	2	2	2-9
Myrica pensylvanica	Northern Bayberry	3	1	2-9
Myrrhis odorata	Sweet Cicely, Anise	4	3	4-8
Nasturtium officinale	Watercress	4	3	3-11
Onoclea sensibilis	Sensitive Fern	2	2	4-8

Plants for the food forest				
Botanical Name	Common Name	Edible Rating	Medicinal Rating	USDA Hardiness Zone
Opuntia compressa	Eastern Prickly Pear, Prickly Pear Cactus	3	1	8-10
Origanum vulgare hirtum	Greek Oregano	4	3	4-8
Osmorhiza claytonii	Woolly Sweet-Cicely, Clayton's sweetroot	3	1	5-9
Osmorhiza longistylis	Aniseroot, Longstyle sweetroot	3	1	5-9
Oxalis montana	Mountain Wood Sorrel	2	0	3-9
Oxalis oregana	Redwood Sorrel	3	1	6-9
Oxalis violacea	Violet Wood Sorrel	3	1	4-8
Oxyria digyna	Mountain Sorrel, Alpine mountain sorrel	3	1	2-8
Panax quinquefolius	American Ginseng	1	3	3-7
Panax trifolius	Ground Nut, Dwarf ginseng	1	2	3-7
Parthenium integrifolium	Wild Quinine	0	2	3-7
Passiflora incarnata	Maypops—Passion Flower, Purple passionflower, Apricot Vine, Maypop, Wild Passion Flower, Purple Pa	3	3	7-11
Perideridia gairdneri	Yampa, Gardner's yampah, Common yampah	5	2	5-10
Persea borbonia	Red Bay, Sweetbay	1	2	7-11
Petasites japonicus	Sweet Coltsfoot, Japanese sweet coltsfoot, Butterbur	3	2	5-9
Phaseolus polystachios	Thicket Bean. Wild bean	2	0	6-10
Phyllostachys dulcis	Sweetshoot Bamboo	4	0	7-10
Phyllostachys nuda		4	0	7-10
Phyllostachys viridiglaucescens	Greenwax golden bamboo	4	0	6-9

Plants for the food forest				
Botanical Name	Common Name	Edible Rating	Medicinal Rating	USDA Hardiness Zone
Physalis heterophylla	Clammy Ground Cherry, Rowell's groundcherry	3	1	7-10
Phytolacca americana	Pokeweed, American pokeweed, Garnet, Pigeon Berry, Poke	3	3	4-8
Pimpinella saxifraga	Burnet Saxifrage, Solidstem burnet saxifrage	2	2	4-8
Pinus albicaulis	White-Bark Pine	4	2	4-8
Pinus armandii	Chinese White Pine, Armand pine	4	2	6-9
Pinus cembra	Swiss Stone Pine, Swiss Pine, Arolla Pine	4	2	3-9
Pinus cembra sibirica	Siberian Pine	4	2	1-6
Pinus edulis	Rocky Mountain Piñon, Twoneedle pinyon, Nut Pine, Pinyon Pine, Rocky Mountain Pinyon Pine, Singlelea	4	2	5-8
Pinus flexilis	Limber Pine, Rocky Mountain White Pine	3	2	4-7
Pinus jeffreyi	Jeffrey Pine	3	2	5-8
Pinus koraiensis	Korean Nut Pine, Chinese pinenut	4	2	4-7
Pinus pumila	Dwarf Siberian Pine	3	2	3-7
Platycodon grandiflorus	Balloon Flower	2	3	4-9
Podophyllum peltatum	American Mandrake, Mayapple, Ground Lemon, Mandrake, Mayapple	2	4	3-9
Polygonatum biflorum	Small Solomon's Seal	2	1	3-7
Polygonum bistorta	Bistort, Meadow bistort, Snakeweed	3	3	4-7
Polygonum bistortoides	American Bistort	3	1	4-8
Polygonum viviparum	Alpine Bistort	3	1	3-7
Polypodium glycyrrhiza	Licorice Fern	1	2	6-9

Plants for the food forest				
Botanical Name	Common Name	Edible Rating	Medicinal Rating	USDA Hardiness Zone
Polypodium vulgare	Polypody, Adders Fern, Golden Maidenhair Fern, Wall Fern, Common Polypod Fern	2	3	3-6
Polystichum acrostichoides	Christmas Fern	1	2	4-9
Potentilla anserina	Silverweed	3	3	4-8
Prosopis glandulosa	Honeypod mesquite. Glandular mesquite	3	2	8-11
Prunus alleghaniensis	Allegheny Plum, Davis' plum	3	1	4-8
Prunus americana	American Plum, American Wild Plum, Wild Plum	3	2	3-8
Prunus nigra	Canadian Plum	4	1	4-9
Prunus angustifolia	Chickasaw Plum, Watson's plum, Hally Jolivette Cherry	3	1	5-9
Prunus angustifolia watsonii	Sand Plum	4	1	5-9
Prunus armeniaca	Apricot	4	3	5-7
Prunus avium	Wild Cherry, Sweet cherry	4	2	3-7
Prunus cerasifera	Cherry Plum, Myrobalan Plum, Newport Cherry Plum, Pissard Plum	4	1	5-8
Prunus cerasus	Sour Cherry	1	2	3-7
Prunus cerasus frutescens	Bush Sour Cherry	3	1	3-7
Prunus domestica	Plum, European plum	5	2	4-9
Prunus insititia	Damson	5	1	5-9
Prunus dulcis	Almond, Sweet almond	4	3	6-9
Prunus fruticosa	Mongolian Cherry, European dwarf cherry	3	1	4-8
Prunus hortulana	Hog Plum, Hortulan plum	3	1	5-9
Prunus japonica nakai	Japanese Plum	3	3	4-8

Plants for the food forest				
Botanical Name	Common Name	Edible Rating	Medicinal Rating	USDA Hardiness Zone
Prunus maritima	Beach Plum, Graves' plum	4	1	3-7
Prunus munsoniana	Wild Goose Plum	3	1	5-9
Prunus persica	Peach, Flowering Peach, Ornamental Peach, Common Peach	5	3	5-9
Prunus persica nucipersica	Nectarine	4	3	4-8
Prunus pumila	Dwarf American Cherry, Sandcherry, Western sandcherry, Eastern sandcherry, Great Lakes sandcherry	4	1	3-8
Prunus salicina	Japanese Plum	2	1	5-8
Prunus tomentosa	Nanking Cherry	3	1	3-9
Prunus virginiana	Chokecherry, Western chokecherry, Black chokecherry	3	2	2-7
Prunus x gondouinii	Duke Cherry	2	1	4-8
Stellaria jamesiana	tuber starwort	3	0	4-9
Psoralea esculenta	Breadroot, Large Indian breadroot	5	1	4-8
Pycnanthemum flexuosum	Mountain Mint, Appalachian mountain mint	1	2	4-8
Pycnanthemum virginianum	Virginia Mountain Mint	2	2	4-8
Pyrus bretschneideri		4	0	4-8
Pyrus communis	Wild Pear, Common pear	2	1	4-9
Quercus acutissima	Sawthorn Oak	2	2	5-9
Quercus alba	White Oak, Hybrid oak	3	2	3-9
Quercus bicolor	Swamp White Oak	4	2	4-8
Quercus fruticosa		3	2	7-10
Quercus gambelii	Shin Oak, Gambel oak, Rocky Mountain White Oak	3	2	4-8

Plants for the food forest				
Botanical Name	Common Name	Edible Rating	Medicinal Rating	USDA Hardiness Zone
Quercus ilex	Holly Oak, Evergreen Oak	5	2	4-10
Quercus macrocarpa	Burr Oak, Mossy Cup Oak	3	2	3-8
Quercus muehlenbergii	Yellow Chestnut Oak, Chinkapin oak	3	2	4-8
Quercus michauxii	Swamp Chestnut Oak	2	2	5-9
Quercus prinoides	Dwarf Chinkapin Oak	2	2	4-8
Quercus prinus	Rock Chestnut Oak	4	2	4-8
Quercus x hybrid	Burgambel oak	3	2	3-8
Rheum australe	Himalayan Rhubarb	3	3	5-9
Rheum palmatum	Turkey Rhubarb, Chinese Rhubarb—Da Huang, Chinese rhubarb	3	5	6-9
Rheum x cultorum	Rhubarb	4	3	3-7
Rhexia virginica	Deer Grass, Handsome Harry	3	1	3-7
Rhodiola rosea	Rose Root	2	3	2-8
Rhus aromatica	Lemon Sumach, Fragrant sumac	4	2	3-9
Rhus copallina	Dwarf Sumach, Winged sumac, Flameleaf Sumac, Winged Sumac, Shining Sumac	4	2	4-10
Rhus glabra	Smooth Sumach	4	3	3-9
Rhus typhina	Stag's Horn Sumach, Velvet Sumac, Staghorn Sumac	4	2	4-8
Ribes americanum	American Blackcurrant	2	1	3-7
Ribes aureum	Golden Currant	4	1	3-8
Ribes hirtellum	Currant-Gooseberry, Hairystem gooseberry	3	0	4-8
Ribes missouriense	Missouri Gooseberry	3	0	4-8

Plants for the food forest				
Botanical Name	Common Name	Edible Rating	Medicinal Rating	USDA Hardiness Zone
Ribes nigrum	Blackcurrant, European black currant	5	3	4-8
Ribes odoratum	Buffalo Currant	4	1	4-8
Ribes rubrum	Red Currant, Cultivated currant	4	1	4-8
Ribes triste	American Red Currant, Red currant	3	1	2-8
Ribes uva-crispa	Gooseberry, European gooseberry	5	1	4-8
Ribes x culverwellii	Jostaberry	5	0	5-9
Robinia pseudoacacia	Black Locust, Yellow Locust	3	2	4-9
Robinia viscosa	Clammy Locust, Hartweg's locust	0	0	3-7
Rosa carolina	Pasture Rose, Carolina rose	2	1	4-8
Rosa rugosa	Ramanas Rose, Rugosa rose	5	2	3-9
Rosa villosa	Apple Rose	4	1	4-8
Rosmarinus officinalis	Rosemary	2	3	6-11
Rubus allegheniensis	Alleghany Blackberry, Graves' blackberry	3	2	3-7
Rubus chamaemorus	Cloudberry	4	1	2-4
Rubus flagellaris	Northern Dewberry	3	1	3-7
Rubus fruticosus	Blackberry, Shrubby blackberry	5	3	5-9
Rubus idaeus	Raspberry, American red raspberry, Grayleaf red raspberry	5	3	4-8
Rubus strigosus	American Red Raspberry, Grayleaf red raspberry	3	3	3-7
Rubus illecebrosus	Strawberry-Raspberry	3	0	4-8
Rubus occidentalis	Black Raspberry	4	2	3-7
Rubus odoratus	Thimbleberry, Purpleflowering raspberry	3	2	3-7

Plants for the food forest				
Botanical Name	Common Name	Edible Rating	Medicinal Rating	USDA Hardiness Zone
Rubus parviflorus	Thimbleberry	3	3	3-7
Rubus rolfei	'Emerald Carpet' raspberry	2	0	7-9
Rubus spectabilis	Salmonberry	3	2	4-8
Rubus tricolor	Creeping Bramble	3	0	6-9
Rumex acetosa	Sorrel, Garden sorrel	5	3	3-7
Rumex acetosella	Sheeps Sorrel, Common sheep sorrel	4	3	4-8
Rumex scutatus	Buckler-Leaved Sorrel, French sorrel	4	1	5-9
Salvia officinalis	Sage, Kitchen sage, Small Leaf Sage, Garden Sage	4	5	5-10
Sambucus nigra spp canadensis	American Elder	4	3	3-9
Sanguisorba canadensis	American Great Burnet, Canadian burnet	1	0	4-8
Sanguisorba minor	Salad Burnet, Small burnet	4	2	4-8
Sanguisorba officinalis	Great Burnet	2	3	4-8
Sasa kurilensis	Chishima Zasa	4	1	6-9
Sassafras albidum	Sassafras, Common Sassafras	5	3	5-9
Micromeria chamissonis	Yerba Buena	2	2	6-9
Saxifraga stolonifera	Strawberry Saxifrage, Creeping Saxifrage, Strawberry Geranium, Strawberry Begonia	2	2	6-10
Schisandra chinensis	Magnolia Vine, Wu Wei Zi	4	5	4-8
Scorzonera hispanica	Scorzonera	4	0	5-9
Secale cereale	Rye, Cereal rye	4	1	3-7
Sedum album	Small Houseleek, White stonecrop, Sedum, Stonecrop	1	1	6-8
Sedum rupestre	Crooked Yellow Stonecrop	1	0	6-9
Semiarundinaria fastuosa	Narihiradake, Narihira bamboo	5	0	6-9

Plants for the food forest				
Botanical Name	Common Name	Edible Rating	Medicinal Rating	USDA Hardiness Zone
Packera aurea	Golden Groundsel—Life Root, Golden ragwort	0	2	3-7
Shepherdia canadensis	Buffalo Berry, Russet buffaloberry, Canada Buffaloberry	3	2	2-6
Silphium laciniatum	Compass Plant, Robinson's compassplant	2	2	4-8
Silphium perfoliatum	Cup Plant, Rosinweed	0	2	3-7
Sium sisarum	Skirret, Suikerwortel (Netherlands), Crummock (Scotland), Zuckewurzel (Germany)	4	0	4-9
Polymnia uvedalia	Bearsfoot	0	2	5-9
Smilacina racemosa	False Spikenard	4	2	2-8
Smilax herbacea	Carrion Flower, Smooth carrionflower	4	1	4-8
Solidago missouriensis	Prairie Goldenrod, Missouri goldenrod, Tolmie's goldenrod	2	1	6-9
Solidago odora	Sweet Goldenrod, Anisescented goldenrod, Chapman's goldenrod	2	2	3-7
Sophora japonica	Japanese Pagoda Tree, Scholar Tree	2	3	4-9
Sorbus americana	American Mountain Ash	1	2	2-6
Sorbus aucuparia	Mountain Ash, European mountain ash	2	2	3-6
Sorbus decora	Showy Mountain Ash	1	1	2-6
Sorbus domestica	Service Tree	5	0	6-10
Sorghum hybrids	Perennial Sorghum	4	1	7-12
Stachys affinis	Chinese Artichoke, Artichoke betony	4	1	4-8
Stachys hyssopifolia	hyssopleaf hedgenettle	2	0	4-8
Stachys palustris	Marsh Woundwort, Marsh hedgenettle	3	2	4-8
Stellaria media	Chickweed, Common chickweed	2	3	4-11

Plants for the food forest				
Botanical Name	Common Name	Edible Rating	Medicinal Rating	USDA Hardiness Zone
Streptopus amplexifolius	Wild Cucumber, Claspleaf twistedstalk, Tubercle twistedstalk	4	2	4-8
Streptopus roseus	Scootberry, Rosybells	4	2	3-7
Symphytum grandiflorum	Ground Cover Comfrey, Comfrey	0	0	3-9
Symphytum officinale	Comfrey, Common comfrey	3	5	3-9
Symphytum uplandicum	Comfrey	3	5	4-8
Taraxacum officinale	Dandelion—Kukraundha, Kanphool, Common dandelion, Dandelion	4	3	5-9
Thymus vulgaris	Common Thyme, Garden thyme, Wild Thyme	4	3	5-11
Tiarella cordifolia	Foamflower, Heartleaf foamflower, Clumping Foamflower	0	2	3-9
Tilia americana	American Basswood, Carolina basswood, Basswood, American Basswood, American Linden	3	3	3-9
Tilia x europaea	Linden, Common Lime	5	3	3-9
Toona sinensis	Chinese Cedar	3	2	6-11
Tradescantia virginiana	Spiderwort, Virginia spiderwort	2	1	4-9
Trifolium pratense	Red Clover	3	3	5-9
Trifolium repens	White Clover, Dutch Clover, Purple Dutch Clover, Shamrock, White Clover	3	2	4-8
Trillium grandiflorum	White Trillium, Large Flower Trillium, White Trillium, Large Flower Wakerobin, Large Flowered Tril	1	2	4-9
Tripsacum dactyloides	Sesame Grass, Eastern gamagrass, Fakahatchee Grass	2	0	7-11

Plants for the food forest				
Botanical Name	Common Name	Edible Rating	Medicinal Rating	USDA Hardiness Zone
Triteleia grandiflora	Wild Hyacinth, Largeflower triteleia, Howell's triteleia	4	0	4-8
Triticum aestivum	Bread Wheat, Common wheat	4	2	10-12
Ulmus rubra	Slippery Elm	2	5	3-7
Urtica dioica	Stinging Nettle, California nettle	5	5	3-10
Urtica californica	Stinging Nettle, California nettle	2	2	3-10
Uvularia sessilifolia	Bellwort, Sessileleaf bellwort	2	1	4-8
Vaccinium angustifolium	Low Sweet Blueberry, Lowbush blueberry	3	1	2-6
Vaccinium ashei	Rabbiteye Blueberry	2	0	7-10
Vaccinium corymbosum	High-Bush Blueberry, American Blueberry, Swamp Blueberry, Blueberry	4	1	3-8
Vaccinium crassifolium	Creeping Blueberry	3	0	6-9
Vaccinium macrocarpon	American Cranberry, Cranberry	3	1	2-7
Vaccinium vitis-idaea	Cowberry, Lingonberry, Northern mountain cranberry, Cranberry	3	2	3-8
Veronica officinalis	Common Speedwell	1	2	3-7
Viburnum cassinoides	Withe Rod, Appalachian Tea, Witherod Viburnum, Witherod, Wild Raisin Viburnum	3	1	2-8
Viburnum lentago	Sheepberry, Nannyberry, Nannyberry Viburnum	4	1	2-8
Viburnum trilobum	American Cranberry, Highbush Cranberry, Cranberry bush, American Cranberry bush Viburnum	3	1	2-7

Plants for the food forest				
Botanical Name	Common Name	Edible Rating	Medicinal Rating	USDA Hardiness Zone
Vicia americana	American Vetch, Mat vetch	2	1	4-7
Vicia cracca	Tufted Vetch, Bird vetch, Cow vetch	1	1	4-8
Viola canadensis	Canada Violet, Canadian white violet, Creeping root violet	3	1	3-8
Viola labradorica	Labrador Violet, Alpine violet, Johnny Jump-Up, Alpine Violet	3	0	3-8
Viola odorata	Sweet Violet, English Violet, Garden Violet, Sweet Violet, Florist's Violet	5	3	4-8
Viola sororia	Wooly Blue Violet, Common blue violet	3	1	4-8
Vitex agnus-castus	Agnus Castus, Lilac chastetree, Vitex, Chastetree	2	5	7-9
Vitis labrusca	Northern Fox Grape, Fox grape	3	1	4-9
Vitis riparia	Riverbank Grape	3	0	2-6
Vitis rotundifolia	Muscadine Grape, Muscadine, Southern Fox Grape, Scuppernong, Muscadine Grape	4	0	5-9
Vitis vinifera	Grape, Wine grape, Purpleleaf Grape, Common Grape	5	2	6-10
Wisteria floribunda	Japanese Wisteria	2	0	5-9
Wisteria frutescens	American Wisteria	1	0	4-8
Xanthoceras sorbifolium	Yellowhorn	3	0	4-7
Yucca baccata	Spanish Bayonet, Banana yucca, Blue Yucca, Spanish Yucca	4	1	6-11

Plants for the food forest				
Botanical Name	Common Name	Edible Rating	Medicinal Rating	USDA Hardiness Zone
Zanthoxylum americanum	Prickly Ash—Northern, Common pricklyash, Northern Prickly Ash	2	3	3-7
Zanthoxylum piperitum	Japanese Pepper Tree	3	2	5-9
Zizia aurea	Golden Alexanders, Golden zizia	2	1	3-7
Ziziphus jujuba	Jujube	4	3	5-10
Asarum splendens	Chinese Wild Ginger	3	0	5-9
Eurybia divaricata	White wood aster	2	0	4-8
Blephilia ciliata	Downy wood mint	3	1	4-7
Blephilia hirsuta	Hairy wood-mint or hairy pagoda plant	2	0	4-7
Camassia cusickii	Cussick's camas	2	0	3-11
Caragana frutex	Russian pea shrub	0	0	2-7
Carex pensylvanica	Pennsylvania sedge	0	0	4-8
Chrysogonum virginianum	Golden-knee, Green and Gold, or Goldenstar	0	0	5-9
Clinopodium glabellum	Glade calamint	0	0	6-7
Clitoria mariana	Atlantic Pigeonwings, Butterfly Pea	0	0	6-9
Conoclinium coelestinum	Blue mistflower	0	0	6-10
Coreopsis rosea	Pink tickseed	0	0	4-9
Coreopsis verticillata	Whorled tickseed	1	1	3-9
Cymopterus purpurascens	Gamote, Widewing springparsley	4	0	6-9
Cytisus decumbens	Prostrate Broom	0	0	5-8
Dennstaedtia punctilobula	Hay-scented fern	0	0	3-8
Desmodium canadense	Showy tick-trefoil	0	0	3-6
Desmodium glutinosum	Pointed-leaved Ticktrefoil	0	0	3-9
Dicentra eximia	Dwarf bleeding heart, turkey-corn	0	0	3-9

Plants for the food forest				
Botanical Name	Common Name	Edible Rating	Medicinal Rating	USDA Hardiness Zone
Equisetum scirpoides	Dwarf scouring rush	2	2	1-9
Erigeron pulchellus	Robin's plantain, blue spring daisy, hairy fleabane	0	0	4-8
Genista pilosa	Hairy greenweed, silkyleaf broom	0	0	5-8
Genista pilosa procumbens	Creeping broom, Creeping hairy broom	0	0	6-8
Genista sagittalis	Winged Broom, Arrow Broom	0	0	3-8
Gymnocarpium dryopteris	Northern oak fern	0	0	2-7
Halesia tetraptera	Silverbell or Mountain Silverbell	2	0	4-8
Helianthus decapetalus	Thinleaf sunflower	0	0	2-8
Heliopsis helianthoides	False sunflower, Oxeye sunflower	2	0	3-9
Houstonia caerulea	Azure bluet or Quaker ladies	2	0	3-8
Ligusticum canbyi	Osha	3	0	3-6
Oxalis grandis	Great Yellow Woodsorrel	2	1	5-7
Pachysandra procumbens	Allegheny spurge	0	0	5-8
Phlox divaricata	Wild blue phlox	0	0	4-9
Phlox nivalis	Pine phlox, Trailing phlox	0	0	6-9
Phlox stolonifera	Creeping phlox	0	0	4-9
Phlox subulata	Moss phlox	0	0	4-9
Abies grandis	Grand Fir, Giant Fir, Lowland White Fir	2	2	5-6
Acacia dealbata	Mimosa, Silver wattle	2	0	7-10
Acacia decurrens	Green Wattle	2	1	6-9
Acanthus mollis	Bear's Breeches	0	2	7-10
Acer pseudoplatanus	Sycamore, Great Maple, Scottish Maple, Planetree Maple	2	1	4-7
Aegopodium podagraria	Ground Elder, Bishop's goutweed, Goutweed, Ground Elder, Bishop's Weed	3	2	4-9

Plants for the food forest				
Botanical Name	Common Name	Edible Rating	Medicinal Rating	USDA Hardiness Zone
Agastache rugosa	Korean Mint	4	3	7-10
Ajuga reptans	Bugle, Common Bugelweed, Bugleweed, Carpet Bugleweed, Carpetweed, Carpet Bugle	2	3	3-10
Akebia quinata	Akebia, Chocolate vine, Fiveleaf Akebia, Chocolate Vine	4	2	4-8
Akebia trifoliata	Akebia, Threeleaf Akebia	4	2	5-8
Allium moly	Golden Garlic, Ornamental Onion	4	2	3-9
Allium neapolitanum	Daffodil Garlic, White garlic	5	2	7-10
Allium ursinum	Wild Garlic	5	3	4-8
Alnus glutinosa	Alder, European alder, Common Alder, Black Alder	0	3	3-7
Alnus rubra	Red Alder, Oregon Alder	2	2	6-8
Alnus sinuata	Sitka Alder	1	1	2-9
Alnus viridis crispa	American Green Alder	1	2	4-8
Angelica sylvestris	Wild Angelica, Woodland angelica	3	2	4-8
Anthemis tinctoria	Yellow Camomile, Golden chamomile, Dyers' Chamomile, Golden Marguerite	0	1	4-6
Apium graveolens	Wild Celery. Ajmod, Ajwain-ka-patta (Indian)	3	3	5-9
Aquilegia vulgaris	Columbine, European columbine, Granny's Bonnet, European Crowfoot	2	1	3-9
Arachis hypogaea	Peanut	4	2	7-10
Arctium lappa	Great Burdock, Gobo	4	5	3-7

Plants for the food forest				
Botanical Name	Common Name	Edible Rating	Medicinal Rating	USDA Hardiness Zone
Arctium minus	Lesser Burdock	3	5	4-8
Aronia prunifolia	Purple Chokeberry	2	0	4-8
Asarum europaeum	Asarabacca, European Wild Ginger	0	2	4-8
Atriplex canescens	Grey Sage Brush, Fourwing saltbush	4	1	6-9
Barbarea verna	Land Cress, Early yellowrocket	3	0	5-9
Berberis aggregata	Salmon Barberry	3	2	5-9
Berberis angulosa		3	2	5-9
Berberis aristata	Chitra, Indian Barberry or Tree Turmeric	4	3	5-9
Berberis asiatica	Chutro, Rasanjan (Nep); marpyashi (Newa); Daruharidra, Darbi (Sans)	4	3	7-10
Berberis buxifolia	Magellan Barberry	4	2	4-8
Berberis cooperi		3	2	4-8
Berberis georgii	Barberry	3	2	3-7
Berberis lycium		3	3	5-9
Berberis rubrostilla		3	2	5-9
Berberis vulgaris	European Barberry, Common barberry	3	3	3-7
Berberis darwinii	Darwin's Barberry, Darwin's berberis	4	2	7-9
Berberis wilsoniae	Wilson barberry	3	2	6-9
Berberis x carminea		3	2	5-9
Berberis x lologensis		3	2	5-9
Berberis x stenophylla		3	2	4-8
Bergenia crassifolia	Siberian Tea	2	0	3-7
Bergenia ciliata		1	2	6-9
Betula pendula	Silver Birch, European white birch, Common Birch, Warty Birch, European White Birch	3	3	2-6

Plants for the food forest				
Botanical Name	Common Name	Edible Rating	Medicinal Rating	USDA Hardiness Zone
Betula pubescens	White Birch, Downy birch	3	3	2-9
Calendula officinalis	calendula, Pot Marigold	3	5	2-11
Calycanthus occidentalis	Californian Allspice, Western sweetshrub	3	1	8-9
Campanula rapunculus	Rampion	4	0	4-8
Cardamine flexuosa	Wavy Bittercress, Woodland bittercress	2	0	4-8
Cardamine hirsuta	Hairy Bittercress	3	0	4-8
Castanea sativa	Sweet Chestnut, European chestnut	5	2	5-7
Cephalotaxus fortunei	Chinese Plum Yew	5	1	6-9
Cercis siliquastrum	Judas Tree, Redbud	4	0	6-9
Cercis occidentalis	Western Redbud, California Redbud	3	0	5-9
Chaenomeles cathayensis	Chinese Quince	4	2	4-8
Chaenomeles japonica	Dwarf Quince, Maule's quince, Japanese Flowering Quince	3	0	5-8
Chaenomeles x superba	Dwarf Quince, Flowering Quince	3	0	5-8
Chamaecyparis lawsoniana	Lawson Cypress, Port orford cedar, Oregon Cedar, Port Orford Cedar, Lawson's Cypress	0	1	5-7
Chamaecyparis nootkatensis	Nootka Cypress, Nootka Cypress, Yellow Cypress, Alaska Cedar	0	1	4-8
Chrysosplenium alternifolium	Golden Saxifrage, Alternate-leaf golden saxifrage, Iowa golden saxifrage	2	0	4-8
Chrysosplenium oppositifolium	Golden Saxifrage	2	0	4-8

Plants for the food forest				
Botanical Name	Common Name	Edible Rating	Medicinal Rating	USDA Hardiness Zone
Citrus aurantiifolia	Lime, Key Lime, Mexican Lime, Mexican Thornless Key Lime	4	2	10-12
Citrus aurantium	Bitter Orange, Sour orange, Bergamot orange	3	3	8-11
Citrus ichangensis	Ichang Papeda	2	2	7-10
Citrus limon	Lemon	4	5	8-11
Citrus reticulata	Mandarin, Tangerine, Unshu orange, Satsuma Orange, Temple Orange, Tangerine	3	3	9-11
Citrus sinensis	Sweet Orange	4	3	9-11
Citrus x meyeri	Lemon	3	5	8-11
Citrus x paradisi	Grapefruit, Pomelo, Pamplemousse	4	1	9-11
Clematis vitalba	Traveller's Joy, Evergreen clematis	1	2	4-8
Cordyline australis	Cabbage Tree	3	0	7-10
Cornus capitata	Bentham's Cornel	4	1	7-10
Corylus maxima	Filbert, Giant filbert	5	0	4-8
Crambe cordifolia	Flowering sea kale	3	1	5-9
Crambe tatarica	Tartar Bread Plant	3	0	4-8
Crataegus arnoldiana	Arnold Hawthorn	5	2	5-9
Crataegus ellwangeriana	Scarlet Hawthorn	5	2	5-7
Crataegus monogyna	Hawthorn, Oneseed hawthorn	3	5	4-8
Crataegus pedicellata	Scarlet Haw, Scarlet hawthorn	5	2	4-8
Crataegus pinnatifida major	Chinese Haw	4	3	5-9
Crataegus tanacetifolia	Tansy-Leaved Thorn	5	2	6-8
Cryptomeria japonica	Japanese Cedar, Sugi	0	1	5-8
Cynara cardunculus	Cardoon	3	5	5-9
Cytisus scoparius	Broom, Scotch broom, Common Broom	1	3	5-8

Plants for the food forest				
Botanical Name	Common Name	Edible Rating	Medicinal Rating	USDA Hardiness Zone
Dioscorea bulbifera	Aerial Yam, Air Potato	4	2	9-12
Diospyros lotus	Date Plum	5	1	7-9
Drimys lanceolata	Mountain Pepper	3	1	7-10
Drimys winteri	Winter's Bark	3	2	7-10
Duchesnea indica	Mock Strawberry, Indian strawberry	2	2	5-11
Echinacea angustifolia	Echinacea, Blacksamson echinacea, Strigose blacksamson	0	5	3-8
Elaeagnus angustifolia	Oleaster, Russian olive	4	2	2-7
Elaeagnus x ebbingei	Elaeagnus, Ebbing's Silverberry	5	2	5-9
Elaeagnus glabra	Goat nipple	4	2	7-10
Elaeagnus pungens	Elaeagnus, Thorny olive, Thorny Elaeagnus, Oleaster, Silverberry, Silverthorn, Pungent Elaeagnus	5	2	6-10
Elaeagnus umbellata	Autumn Olive	4	2	3-7
Eucalyptus johnstonii	Yellow Gum, Johnston's gum	0	0	7-10
Fagopyrum esculentum	Buckwheat	4	3	6-12
Filipendula ulmaria	Meadowsweet, Queen of the meadow, Double Lady of the Meadow, European Meadowsweet	3	3	3-9
Ficus carica	Fig, Edible fig, Fig Common	4	2	6-10
Fragaria viridis	Green Strawberry	3	0	5-9
Fraxinus excelsior	Ash, European ash, Common Ash	2	2	5-8
Fuchsia magellanica	Fuchsia, Hardy fuchsia	2	1	5-7
Garrya elliptica	Coast Silk Tassel, Wavyleaf silktassel	0	2	7-10
Garrya fremontii	Fever Bush, Bearbrush	0	1	6-9

Plants for the food forest				
Botanical Name	Common Name	Edible Rating	Medicinal Rating	USDA Hardiness Zone
Geranium macrorrhizum	Bigroot Geranium	0	1	4-8
Glechoma hederacea	Ground Ivy, Field Balm, Gill Over The Ground, Runaway Robin	2	3	3-10
Glycyrrhiza echinata	Wild Liquorice, Chinese licorice	4	3	7-10
Gunnera tinctoria	Gunnera, Chilean gunnera	1	1	6-9
Gunnera magellanica		1	0	6-9
Gynostemma pentaphyllum	Sweet Tea Vine	2	5	7-10
Hedera helix	Ivy, English ivy, Algerian ivy, Baltic Ivy, Common Ivy	0	3	5-11
Hemerocallis aurantiaca		4	1	5-9
Hemerocallis bulbiferum	Orange lily	4	1	3-9
Hemerocallis citrina	Citron daylily	4	1	4-8
Hemerocallis coreana		4	1	3-9
Hemerocallis darrowiana	Day Lily	4	1	3-9
Hemerocallis dumortieri	Dumortier's daylily	4	1	4-8
Hemerocallis exaltata		4	1	4-8
Hemerocallis fulva longituba		4	1	4-8
Hemerocallis littorea	Coastal Day Lily	4	1	4-8
Hemerocallis middendorffii	Amur daylily, Middendorf, Daylily	5	1	4-8
Hemerocallis minor	Grassleaf Day Lily, Small daylily	4	1	4-8
Hemerocallis thunbergii		4	1	4-8
Hemerocallis yezoensis		4	1	4-8
Hippophae salicifolia	Willow-Leaved Sea Buckthorn	5	3	3-7
Hosta longipes		3	0	4-8
Hosta crispula		2	0	4-8
Hosta longissima	Swamp Hosta	2	0	4-8

Plants for the food forest				
Botanical Name	Common Name	Edible Rating	Medicinal Rating	USDA Hardiness Zone
Hosta montana		2	0	4-8
Hosta sieboldiana		2	0	4-8
Hosta tardiva	Nankai-Giboshi	2	0	4-8
Hosta undulata		2	0	4-8
Hosta ventricosa	Blue plantain lily, Hosta	2	1	3-9
Hydrastis canadensis	Goldenseal	0	3	3-7
Hypericum perforatum	St. John's Wort, Common St. Johnswort	2	4	3-7
Juniperus communis	Juniper, Common juniper	3	3	4-10
Larix decidua	Larch, European Larch, Common Larch	2	3	3-6
Larix kaempferi	Japanese Larch	0	0	4-6
Lathyrus odoratus	Sweet Pea, Wild Pea, Vetchling	1	0	2-11
Laurus nobilis	Bay Tree, Sweet bay, Grecian Laurel, True Laurel,	3	3	8-10
Lavandula angustifolia	English Lavender, True Lavender	2	3	5-8
Lavandula x intermedia	Lavender, Lavandin	2	2	5-9
Lavandula latifolia	Spike Lavender, Broadleaved lavender	2	2	3-9
Lavandula stoechas	French Lavender	0	2	7-10
Lespedeza maximowiczii		0	0	4-8
Limnanthes douglasii	Poached Egg Plant, Douglas' meadowfoam, Ornduff's meadowfoam	0	0	4-8
Linaria vulgaris	Yellow Toadflax, Butter and eggs	1	2	4-8
Lolium perenne	Perennial Ryegrass, Italian ryegrass, Darnel, Lyme Grass, Terrell Grass, English Ryegrass, Strand Wh	1	1	5-7

Plants for the food forest				
Botanical Name	Common Name	Edible Rating	Medicinal Rating	USDA Hardiness Zone
Lonicera periclymenum	Honeysuckle, European honeysuckle	1	2	4-8
Lonicera caerulea	Sweetberry honeysuckle, Bluefly honeysuckle, Haskap berry	4	0	3-9
Lotus corniculatus	Bird's Foot Trefoil	1	1	3-8
Lotus uliginosus	Greater Bird's Foot Trefoil	0	0	5-9
Lunaria annua	Honesty, Annual honesty, Silver Dollar, Moneywort, Moonwort, Penny Flower, Money Plant	2	0	8-10
Lupinus angustifolius	Blue Lupin, Narrowleaf lupine	4	0	7-9
Lupinus arboreus	Tree Lupin, Yellow bush lupine	0	0	7-10
Lupinus perennis	Sundial Lupine	3	1	4-8
Lycium barbarum	Goji, Box Thorn, Matrimony vine	4	3	6-9
Lysimachia nummularia	Creeping Jenny, Moneywort, Creeping Charlie	1	2	4-8
Mahonia nervosa	Oregon Grape, Cascade barberry	3	2	5-9
Malus domestica	Apple	5	2	3-8
Malva pusilla	Dwarf Mallow, Low mallow	4	2	5-9
Malva sylvestris	Mallow, High mallow, French Hollyhock, Common Mallow, Tree Mallow, Tall Mallow	3	3	4-8
Malva verticillata	Chinese Mallow, Cluster mallow	5	2	6-12

Plants for the food forest				
Botanical Name	Common Name	Edible Rating	Medicinal Rating	USDA Hardiness Zone
Mentha aquatica	Water Mint	3	3	5-9
Mentha longifolia	Horsemint	2	2	5-9
Myrica californica	Californian Bayberry, California Wax Myrtle, California Barberry	3	1	7-11
Myrica rubra	Chinese Bayberry	2	2	9-11
Myrtus communis	Myrtle, Foxtail Myrtle	3	3	9-11
Ugni molinae	Uñi, Chilean guava	5	0	7-11
Onobrychis viciifolia	Sainfoin	1	0	5-9
Ononis spinosa	Spiny Rest Harrow	2	2	5-9
Origanum vulgare	Oregano, Pot Marjoram	4	3	4-10
Oxalis acetosella	Wood Sorrel	3	2	3-7
Oxalis tuberosa	Oca	5	0	6-9
Panax ginseng	Ginseng, Chinese ginseng	2	5	5-9
Parthenocissus quinquefolia	Virginia Creeper, Woodbine	2	2	3-10
Parthenocissus tricuspidata	Boston Ivy, Japanese Ivy, Japanese Creeper	1	0	4-8
Peltaria alliacea	Garlic Cress	4	0	5-9
Petroselinum crispum	Parsley	4	4	4-9
Phacelia tanacetifolia	Fiddleneck, Lacy phacelia	0	0	3-10
Philadelphus lewisii	Mock Orange, Lewis' mock orange	0	1	4-8
Phleum pratense	Timothy	0	1	4-8
Phormium tenax	New Zealand Flax, Coastal Flax, New Zealand Hemp	2	0	8-10
Phormium cookianum	Wharariki	2	0	7-10
Phyllostachys angusta	Stone Bamboo	3	0	7-10
Phyllostachys aurea	Golden Bamboo, Fishpole Bamboo	5	0	6-11

Plants for the food forest				
Botanical Name	Common Name	Edible Rating	Medicinal Rating	USDA Hardiness Zone
Phyllostachys aureosulcata	Yellow-Groove Bamboo	4	0	5-11
Phyllostachys bambusoides	Madake, Japanese timber bamboo	4	1	6-9
Phyllostachys bissetii		0	0	4-8
Phyllostachys edulis	Moso-Chiku, Tortoise shell bamboo	4	1	6-10
Phyllostachys flexuosa	Zig-Zag Bamboo, Drooping timber bamboo	3	0	5-9
Phyllostachys glauca	Blue bamboo	3	0	7-10
Phyllostachys nidularia	Big-Node Bamboo, Broom bamboo	5	0	6-9
Phyllostachys nigra	Black Bamboo, Kuro-Chiku	4	3	7-10
Phyllostachys rubromarginata	Reddish bamboo	3	0	7-10
Phyllostachys sulphurea viridis	Kou-Chiku	4	0	6-9
Phyllostachys vivax	Giant Timber Bamboo, Running giant bamboo	3	0	7-10
Picea abies	Norway Spruce	2	1	2-7
Picea sitchensis	Sitka Spruce	2	2	6-7
Pinus cembroides	Mexican Pine Nut, Pinyon Pine	4	2	5-8
Pinus gerardiana	Chilghoza Pine	3	2	6-9
Pinus monophylla	Single Leaf Piñon, Single Leaf PinyonPine, Stone Pine, Pine Pinyon	4	2	6-8
Pinus nigra laricio	Corsican Pine	1	2	5-9
Pinus pinea	Italian Stone Pine, Umbrella Pine, Stone Pine	4	2	7-11
Pinus radiata	Monterey Pine	1	2	3-11
Pinus sylvestris	Scot's Pine, Scotch Pine	2	3	3-7

Plants for the food forest				
Botanical Name	Common Name	Edible Rating	Medicinal Rating	USDA Hardiness Zone
Plantago major	Common Plantain, Cart Track Plant, White Man's Foot, Plantain	2	3	3-12
Plantago lanceolata	Ribwort Plantain, Narrowleaf plantain	2	3	5-9
Pleioblastus simonii	Medake, Simon bamboo	3	0	5-9
Polygonatum multiflorum	Solomon's Seal, Eurasian Solomon's seal	2	3	4-8
Polygonatum commutatum	King Solomon's Seal, Smooth Solomon's seal	2	1	4-8
Polygonatum odoratum	Solomon's Seal	2	3	4-8
Poncirus trifoliata	Bitter Orange, Hardy orange, Trifoliat Orange, Japanese Hardy Orange	3	2	6-9
Populus x canadensis	Canadian Poplar, Carolina Poplar	0	1	4-9
Populus nigra	Black Poplar, Lombardy poplar	1	3	4-9
Primula vulgaris	Primrose, Common Primrose, English Primrose	3	3	3-9
Prunus spinosa	Sloe—Blackthorn	3	2	4-8
Pseudosasa japonica	Metake—Bamboo	2	1	5-9
Pseudotsuga menziesii	Douglas Fir, Rocky Mountain Douglas-fir	2	2	3-6
Pulmonaria officinalis	Lungwort, Common lungwort, Jerusalem Sage, Jerusalem Cowslip	2	3	6-9
Pyrus pyrifolia	Sand Pear, Chinese pear	4	1	5-9
Pyrus ussuriensis	Harbin Pear, Chinese pear, Ussurian Pear	4	0	3-7
Quercus douglasii	Blue Oak	3	2	8-11
Quercus ilex ballota	Holm Oak	5	2	6-9

Plants for the food forest				
Botanical Name	Common Name	Edible Rating	Medicinal Rating	USDA Hardiness Zone
Quercus kelloggii	Californian Black Oak, Black Oak	3	2	8-11
Quercus ithaburensis macrolepis	Valonia Oak	4	2	6-9
Quercus petraea	Sessile Oak, Durmast oak	2	3	5-8
Quercus robur	Pedunculate Oak, English oak	4	3	4-8
Raphanus sativus	Radish, Cultivated radish	4	3	2-11
Reichardia picroides	French Scorzonera, Common brighteyes	5	0	7-10
Rheum nobile	Sikkim Rhubarb	3	2	6-9
Rheum officinale	Chinese Rhubarb	1	3	6-9
Rheum palmatum tanguticum	Da Huang	3	5	5-9
Ribes divaricatum	Coastal Black Gooseberry, Spreading gooseberry, Parish's gooseberry, Straggly gooseberry	4	1	4-8
Rosa banksiae	Banksia Rose	2	1	6-9
Rubus loganobaccus	Loganberry	5	0	7-10
Rubus nepalensis	Nepalese Raspberry	5	0	7-10
Rubus phoenicolasius	Japanese Wineberry, Wine raspberry	5	0	4-8
Rumex alpinus	Alpine Dock, Munk's rhubarb	4	2	4-8
Rumex patientia	Herb Patience	3	1	5-10
Salix purpurea	Purple Osier, purpleosier willow	1	3	4-8
Salix triandra	Almond-Leaved Willow, Almond willow	1	2	4-8
Salix viminalis	Osier. Basket Willow	1	2	4-8

Plants for the food forest				
Botanical Name	Common Name	Edible Rating	Medicinal Rating	USDA Hardiness Zone
Sambucus nigra	Elderberry—European Elder, Black elderberry, American black elderberry, Blue elderberry, Europea	4	3	5-7
Sambucus racemosa	Red Elder, Red elderberry, Rocky Mountain elder, European Red Elderberry	3	2	4-8
Sanguinaria canadensis	Blood Root, Red Puccoon, Bloodroot	0	3	4-9
Saponaria officinalis	Soapwort, Bouncingbet	0	3	4-8
Sedum spectabile	Ice Plant	2	1	5-9
Shepherdia argentea	Buffalo Berry, Silver Buffaloberry,	3	1	3-9
Sinapis alba	White Mustard	3	3	5-9
Sisymbrium officinale	Hedge Mustard	1	2	5-9
Polymnia edulis	Yacon Strawberry	4	0	7-10
Smyrnium olusatrum	Alexanders	3	1	5-9
Solidago canadensis	Canadian Goldenrod, Shorthair goldenrod, Harger's goldenrod, Rough Canada goldenrod, Common Goldenro	2	2	5-10
Solidago virgaurea	Goldenrod	1	3	4-8
Sorbus aria	Whitebeam, Chess-apple	3	1	4-8
Sorbus devoniensis	Devon Whitebeam	3	0	6-9
Sorbus thibetica	Tibetan whitebeam	3	0	5-9
Sorbus torminalis	Wild Service Tree, Checkertree	4	0	5-9

Plants for the food forest				
Botanical Name	Common Name	Edible Rating	Medicinal Rating	USDA Hardiness Zone
Symphoricarpos orbiculatus	Coralberry	1	1	2-7
Symphytum tuberosum	Tuberous comfrey	2	0	4-8
Tanacetum parthenium	Feverfew, Matricaria	2	5	5-8
Tanacetum vulgare	Tansy, Common tansy, Golden Buttons, Curly Leaf Tansy	2	2	3-9
Taxus baccata	Yew, English yew, Common Yew	3	4	5-7
Taxus brevifolia	Pacific Yew	3	4	5-9
Taxus canadensis	Canadian Yew	3	4	4-8
Taxus cuspidata	Japanese Yew	3	4	4-7
Taxus x media	Anglojapanese Yew	3	4	4-7
Thuja plicata	Western Red Cedar, Giant Arborvitae, Giant Cedar, Incense Cedar, Western Red Cedar	1	2	5-8
Tilia cordata	Small Leaved Lime, Littleleaf linden	5	3	3-7
Tilia platyphyllos	Large Leaved Lime, Largeleaf linden, Bigleaf Linden	5	3	4-6
Tilia tomentosa	Silver Lime	3	1	4-7
Trifolium incarnatum	Crimson Clover	2	0	3-9
Trifolium subterraneum	Subterranean Clover	1	0	7-10
Tropaeolum majus	Nasturtium, Indian Cress	4	3	8-11
Tropaeolum tuberosum	Anu	4	2	7-10
Tsuga canadensis	Canadian Hemlock, Eastern hemlock	1	3	4-7
Ulex europaeus	Gorse, Common gorse	1	1	5-9
Ullucus tuberosus	Olluco	3	0	8-10
Vaccinium arctostaphylos	Caucasian Whortleberry	3	0	5-9
Vaccinium darrowii	Darrow's blueberry	1	0	8-10
Vaccinium oxycoccos	Small Cranberry	4	1	3-6

Plants for the food forest				
Botanical Name	Common Name	Edible Rating	Medicinal Rating	USDA Hardiness Zone
Valeriana officinalis	Valerian, Garden valerian	2	3	4-8
Vicia faba equina	Horsebean	3	0	4-8
Vicia faba minuta	Tick Bean	3	0	4-8
Vicia sativa	Winter Tares, Garden vetch, Subterranean vetch	3	0	4-8
Vinca major	Greater Periwinkle, Bigleaf periwinkle, Myrtle, Large Periwinkle, Big Periwinkle	0	3	7-9
Vinca minor	Lesser Periwinkle, Flower of Death, English Holly, Creeping Myrtle, Creeping Vinca, Common Periwink	0	3	4-9
Wisteria sinensis	Chinese Wisteria	1	1	5-9
Yushania anceps	Ringal	0	0	8-11
Yushania maling		0	0	8-11
Zanthoxylum alatum	Winged Prickly Ash	3	2	5-9
Zanthoxylum schinifolium	Peppertree	2	2	5-9
Zanthoxylum simulans	Szechuan Pepper, Chinese-pepper, Prickly Ash	3	2	5-8
Salix 'Bowles hybrid'		1	2	8-11
Phyllostachys praecox	Violet Bamboo	3	0	6-10
Stylosanthes biflora	Sidebeak pencilflower	0	0	4-8
Strophostyles umbellata	Pink fuzzybean, Perennial wild bean	0	0	6-9
Robinia hispida	Bristly locust, Rose-acacia, or Moss locust	0	1	4-8
Thermopsis villosa	Aaron's rod, Carolina lupine	0	0	5-8
Rhamnus frangula	Alder Buckthorn	0	3	3-7
Aloe vera	Aloe Vera, Barbados aloe, First Aid Plant, Medicinal Aloe	1	5	9-11

Plants for the food forest				
Botanical Name	Common Name	Edible Rating	Medicinal Rating	USDA Hardiness Zone
Amaranthus caudatus	Love Lies Bleeding	4	1	4-8
Amaranthus blitum	Slender Amaranth, Purple amaranth	4	2	4-8
Amaranthus cruentus	Purple Amaranth, Red amaranth	4	2	3-11
Amaranthus hybridus	Rough Pigweed, Slim amaranth	4	1	6-12
Amaranthus hypochondriacus	Prince's Feather, Prince-of-wales feather	4	3	3-10
Amaranthus retroflexus	Pigweed, Redroot amaranth, Wild Beet	3	2	3-11
Amaranthus tricolor	Chinese Spinach, Joseph's-coat, Fountain Plant, Tampala, Summer Poinsettia	3	1	3-11
Aster macrophyllus	Bigleaf Aster	2	1	3-7
Aster cordifolius	Common Blue Wood Aster	2	1	3-7
Aster lanceolatus	White Panicle Aster	0	1	4-8
Aster novi-belgii	Michaelmas Daisy, New York Aster	0	0	4-8
Aster puniceus	Purplestem Aster	0	2	3-8
Aster scaber		3	0	6-9
Cedrus atlantica	Atlas Deodar, Atlantic cedar	0	2	5-9
Persea americana	Avocado, Alligator Pear	5	3	9-12
Allium ampeloprasum babingtonii	Babington's Leek	3	3	5-9
Waldsteinia fragarioides	Appalachian barren strawberry	0	2	4-7
Erica x darleyensis	Darley Dale Heath, Cape Heath, Molten Silver Heath, Heather	0	0	6-9
Erica vagans	Cornish Heath, Cornish heath	0	0	4-8

Plants for the food forest				
Botanical Name	Common Name	Edible Rating	Medicinal Rating	USDA Hardiness Zone
Erica tetralix	Bog Heather, Crossleaf heath	0	0	3-7
Mentha x piperita citrata	Eau De Cologne Mint, Eau de Cologne Mint, Peppermint	2	2	3-9
Vaccinium myrtillus	Bilberry, Whortleberry	4	3	3-7
Prunus padus	Bird Cherry, European bird cherry	3	2	3-6
Medicago lupulina	Black Medick	2	1	4-8
Fagopyrum dibotrys	Perennial Buckwheat	4	2	7-10
Daucus carota sativus	Carrot	5	3	4-10
Casuarina cristata	Belah	0	0	8-11
Casuarina cunninghamiana	River She-Oak	0	0	8-11
Casuarina glauca	Swamp Oak, Gray sheoak	1	0	8-11
Casuarina littoralis	She Oak, Black she-oak	0	0	9-11
Casuarina torulosa	Forest Oak	0	0	8-11
Casuarina verticillata	Drooping she-oak	0	0	8-11
Tussilago farfara	Coltsfoot	3	3	4-8
Cotoneaster divaricatus	Spreading Cotoneaster	0	0	4-8
Cotoneaster microphyllus		2	1	4-8
Cotoneaster simonsii	Simons' cotoneaster	0	0	4-8
Cotoneaster racemiflorus	Black-Wood, Cotoneaster	2	1	3-7
Malus sieversii	Crabapple	3	0	4-10
Ranunculus repens	Creeping Buttercup, Prairie Double-flowered Buttercup, Water Buttercup, Creeping Buttercup	1	1	3-8
Brassica oleracea ramosa	Perpetual Kale	4	0	6-9
Lamium album	White Dead Nettle	2	3	5-9

Plants for the food forest				
Botanical Name	Common Name	Edible Rating	Medicinal Rating	USDA Hardiness Zone
Lamium galeobdolon	Yellow Archangel	2	1	3-9
Rumex crispus	Curled Dock, Curly dock	2	3	4-8
Inula helenium	Elecampane, Elecampane inula	3	3	4-8
Chenopodium ambrosioides	Mexican Tea	2	3	7-10
Oenothera odorata		1	0	4-8
Oenothera elata hookeri	Hooker's Evening Primrose	2	1	6-9
Oenothera biennis	Evening Primrose, Sun Drop, Common evening primrose	3	5	4-8
Chenopodium album	Fat Hen, Lambsquarters	3	2	1-12
Iris pseudacorus	Yellow Flag, Paleyellow iris	1	2	5-8
Forsythia suspensa	Lian Qiao, Weeping forsythia	1	3	5-8
Forsythia x intermedia	Golden Bell, Border Forsythia	0	0	5-8
Forsythia viridissima	Golden Bells, Greenstem forsythia, Forsythia	0	2	5-8
Digitalis purpurea	Foxglove, Purple foxglove, Common Foxglove	0	4	4-8
Allium sativum	Garlic, Cultivated garlic	5	5	7-10
Cynara scolymus	Globe Artichoke	3	5	5-9
Muscari comosum	Tassel Hyacinth, Tassel grape hyacinth	3	1	4-8
Muscari botryoides	Italian Grape Hyacinth, Common grape hyacinth, White Grape Hyacinth	1	0	3-8
Muscari neglectum	Grape Hyacinth, Starch grape hyacinth	2	0	4-7

Plants for the food forest				
Botanical Name	Common Name	Edible Rating	Medicinal Rating	USDA Hardiness Zone
Verbascum thapsus	Great Mullein, Common mullein, Aaron's Rod, Flannel Plant, Hag Taper, Mullein, Torches, Velvet Plant	1	3	3-8
Pentaglottis sempervirens	Evergreen bugloss	1	0	6-9
Viburnum opulus	Guelder Rose, Cramp Bark, European cranberrybush, American cranberrybush, Crampbark, European Highb	3	3	3-8
Hablitzia tamnoides	Caucasian spinach	4	0	3-9
Vicia hirsuta	Hairy Tare, Tiny vetch	2	0	4-9
Juglans ailanthifolia cordiformis	Heartseed Walnut	4	1	4-8
Juglans ailanthifolia	Japanese Walnut	3	1	4-8
Calluna vulgaris	Heather, Scotch Heather	2	2	4-7
Hebe speciosa	New Zealand hebe	0	0	6-9
Hebe dieffenbachii		0	0	8-11
Hebe rakaiensis		0	0	5-9
Hebe salicifolia		0	1	6-9
Hebe x franciscana	Hebe	0	0	9-11
Alliaria petiolata	Garlic Mustard	3	2	6-8
Geranium robertianum	Herb Robert, Robert geranium	0	2	5-9
Ilex aquifolium	Holly, English holl, Christmas Holly, Common Holly, English Holly	2	2	5-9
Marrubium vulgare	White Horehound, Horehound	1	3	3-7
Equisetum arvense	Field Horsetail	2	3	3-11

Plants for the food forest				
Botanical Name	Common Name	Edible Rating	Medicinal Rating	USDA Hardiness Zone
Equisetum hyemale	Dutch Rush, Scouringrush horsetail, Horsetail, Scouring Rush, Rough Horsetail	2	2	3-11
Hyssopus officinalis	Hyssop	2	3	5-10
Reynoutria japonica	Japanese knotweed.	3	3	4-8
Eupatorium purpureum	Gravel Root	1	3	3-9
Juniperus communis nana	Juniper	3	3	4-10
Juniperus conferta	Shore Juniper	2	0	6-10
Juniperus excelsa	Grecian Juniper	2	1	5-9
Juniperus horizontalis	Creeping Juniper, Horizontal Juniper	2	1	4-9
Juniperus monosperma	One-Seed Juniper	3	2	4-8
Juniperus occidentalis	Western Juniper	3	2	4-8
Juniperus osteosperma	Desert Juniper, Utah juniper	2	2	4-8
Juniperus sabina	Savine, Tam Juniper	0	2	4-7
Juniperus scopulorum	Rocky Mountain Juniper, Weeping Rocky Mountian Juniper, Colorado Red Cedar	3	2	3-7
Juniperus silicicola	Southern Redcedar, Juniper, Southern Red Cedar	2	2	7-10
Juniperus virginiana	Pencil Cedar, Eastern redcedar, Southern redcedar, Silver Cedar, Burk Eastern Red Cedar, Silver East	2	2	3-9
Actinidia chinensis	Kiwi	4	2	6-9
Laburnum anagyroides	Laburnum, Golden chain tree	0	1	4-8
Aloysia citriodora	Lemon Verbena, Lemon beebrush	4	3	7-10
Lactuca sativa	Lettuce, Garden lettuce	3	3	5-9
Liquidambar styraciflua	Sweet Gum, Red Gum, American Sweet Gum, Red Sweet Gum,	2	3	5-9
Eriobotrya japonica	Loquat, Japanese Loquat	4	3	8-11

Plants for the food forest				
Botanical Name	Common Name	Edible Rating	Medicinal Rating	USDA Hardiness Zone
Origanum majorana	Sweet Marjoram	3	3	6-9
Tagetes erecta	African Marigold, Aztec marigold, Big Marigold, American Marigold	3	3	2-11
Viscum album	Mistletoe, European mistletoe	1	3	6-9
Leonurus cardiaca	Motherwort, Common motherwort	2	3	3-7
Verbascum thapsus	Great Mullein, Common mullein, Aaron's Rod, Flannel Plant, Hag Taper, Mullein, Torches, Velvet Plant	1	3	3-8
Verbascum nigrum	Dark Mullein, Black mullein	0	3	4-8
Melaleuca alternifolia	Tea Tree	0	5	8-11
Nerium oleander	Oleander, Rose Bay	0	2	9-11
Atriplex hortensis	Orach, Garden orache	4	2	5-9
Leucanthemum vulgare	Ox-Eye Daisy, Marguerite	2	2	3-9
Pastinaca sativa	Parsnip, Wild parsnip	4	1	4-8
Pyrus communis sativa	Pear	5	0	4-8
Mentha pulegium	Pennyroyal	3	3	6-9
Punica granatum	Pomegranate, Dwarf Pomegranate	3	3	8-12
Papaver rhoeas	Corn Poppy, Field Poppy, Shirley Poppy	2	3	4-8
Solanum tuberosum	Potato, Irish potato	5	2	8-9
Cucurbita pepo	Pumpkin, Field pumpkin, Ozark melon, Texas gourd	4	3	2-11
Lythrum salicaria	Purple Loosestrife	2	3	3-10
Tanacetum coccineum	Pyrethrum, Pyrethum daisy, Persian Insect Flower, Painted Daisy	0	0	4-10

Plants for the food forest				
Botanical Name	Common Name	Edible Rating	Medicinal Rating	USDA Hardiness Zone
Chenopodium quinoa	Quinoa, Goosefoot, Pigweed, Inca Wheat	5	0	10-12
Silene dioica	Red Campion, Red catchfly	0	0	5-9
Eruca vesicaria sativa	Rocket	4	1	6-9
Phaseolus coccineus	Runner Bean, Scarlet runner	4	0	1-2
Perilla frutescens	Shiso, Beefsteakplant, Spreading Beefsteak Plant	4	3	7-10
Scutellaria lateriflora	Virginian Skullcap, Blue skullcap	0	3	6-9
Antirrhinum majus	Snapdragon, Garden snapdragon	1	1	5-10
Sonchus arvensis	Field Milk Thistle, Field sowthistle, Moist sowthistle	2	1	3-8
Sonchus oleraceus	Sow Thistle, Common sowthistle	2	2	7-9
Spinacia oleracea	Spinach	3	2	4-8
Euonymus europaeus	Spindle Tree, European spindletree	1	2	3-7
Hypericum perforatum	St. John's Wort, Common St. Johnswort	2	4	3-7
Trachelospermum jasminoides	Star Jasmine, Confederate jasmine	0	2	8-10
Portulaca oleracea	Green Purslane, Little hogweed	4	3	3-12
Helianthus annuus	Sunflower, Common sunflower	5	2	6-9
Zea mays	Sweetcorn, Corn	5	3	2-11
Cirsium arvense	Creeping Thistle, Canada thistle	2	2	4-7
Cirsium japonicum	No-Azami, Japanese thistle	2	2	5-9

Plants for the food forest				
Botanical Name	Common Name	Edible Rating	Medicinal Rating	USDA Hardiness Zone
Cirsium vulgare	Common Thistle, Bull thistle, Dodder, Boar Thistle, Bull Thistle	2	1	2-10
Zanthoxylum planispinum	Winged Prickly Ash	3	2	5-9
Tulipa edulis		2	2	6-9
Ocimum tenuiflorum	Sacred Basil	3	4	10-12
Viola mirabilis	wonder violet	3	1	4-8
Viola cucullata	Marsh Blue Violet	3	1	3-8
Viola pedata	Bird's Foot Violet, Crowfoot Violet, Pansy Violet, Bird's Foot Violet	2	1	4-9
Viola riviniana	Wood Violet	3	0	4-8
Daucus carota	Wild Carrot, Queen anne's lace, Carrot, Wild Carrot, Queen Anne's Lace	2	3	4-8
Stachys officinalis	Wood Betony, Common hedgenettle, Betony, Woundwort	1	2	5-10
Lamium galeobdolon	Yellow Archangel	2	1	3-9
Capsicum annuum	Sweet Pepper, Cayenne Pepper, Chili Pepper, Christmas Pepper, Red Pepper, Ornamental Chili Pepper	4	3	10-11
Epilobium hirsutum	Codlins And Cream	2	1	2-9
Stellaria graminea	Lesser stitchwort	3	1	4-8
Carex eburnea	Bristleleaf sedge	0	0	2-8
Prunus domestica italica	Gages, Greengage	5	2	5-8
Carex hachijoensis	Japanese sedge	0	0	5-9
Acmella oleracea	Toothache plant, Paracress	3	3	9-11
Persicaria odorata	Vietnamese coriander, Asian mint	3	2	9-11
Carex sylvatica	Wood sedge	0	0	3-8
Murraya koenigii	Curry tree, Curry leaf tree	3	3	10-12
Solanum torvum	Pea Eggplant, Turkey berry	3	3	8-11
Gaillardia x grandiflora	Blanket flower	1	0	3-10

Table 5—Top 100 nitrogen fixers for temperate climates

Sun
⊕ Partial shade
o Sun
● Shade

Soil
D = dry
M = moist
W = wet

Botanical Name	Common Name	USDA Hardiness Zones	Sun	Soil
Acacia angustissima	Prairie acacia. Timbre. Fernleaf Acacia	7-10	o	D to M
Acacia dealbata	Mimosa, Silver wattle	7-10	o	D to M
Acacia decurrens	Green Wattle	6-9	o	D to M
Alnus cordata	Italian Alder	5-9	⊕ o	D,M,W
Alnus glutinosa	Alder, European alder, Common Alder, Black Alder	3-7	⊕ o	M, W
Alnus incana	Grey Alder, Speckled alder, Thinleaf alder, White Alder	2-6	⊕ o	D,M,W
Alnus rubra	Red Alder, Oregon Alder	7-8	⊕ o	M,W
Alnus rugosa	Speckled Alder	2-6	⊕ o	M,W
Alnus serrulata	Smooth Alder, Hazel alder	3-9	o	M,W
Alnus sinuata	Sitka Alder	3-9	⊕ o	M,W

Botanical Name	Common Name	USDA Hardiness Zones	Sun	Soil
Alnus viridis crispa	American Green Alder	4-8	⊕ ○	M,W
Amorpha canescens	Lead Plant	2-9	⊕ ○	D to M
Amorpha fruticosa	False Indigo, False indigo bush	4-8	⊕ ○	D to M
Amorpha nana	Dwarf Indigobush, Dwarf false indigo, Dwarf Indigo	4-8	⊕ ○	D to M
Amphicarpaea bracteata	Hog Peanut, American hogpeanut	4-9	⊕ ●	M
Apios americana	Ground Nut	3-7	⊕ ○	M
Apios fortunei	Hodo, Hodoimo	4-9	⊕ ○	M
Apios priceana	Traveler's delight	6-9	⊕ ○	M
Arachis hypogaea	Peanut	7-10	○	M
Astragalus canadensis	Canadian Milkvetch, Shorttooth Canadian milkvetch, Morton's Canadian milkvetch	7-10	○	D
Astragalus crassicarpus	Ground Plum, Groundplum milkvetch	6-9	○	D
Astragalus glycyphyllos	Milk Vetch, Licorice milkvetch	3-7	○	D

Botanical Name	Common Name	USDA Hardiness Zones	Sun	Soil
Astragalus membranaceus	Huang Qi	5-9	○	D
Baptisia australis	Wild Indigo, Blue wild indigo, Blue False Indigo	3-9	○	D to M
Baptisia tinctoria	Wild Indigo, Horseflyweed	4-8	○	D to M
Caragana arborescens	Siberian Pea Tree, Siberian peashrub	2-7	○	D to M
Caragana pygmaea	Caragana pygmaea	3-7	○	D to M
Ceanothus americanus	New Jersey Tea, Wild Snowball	4-9	⊕ ○	D to M
Ceanothus prostratus	Squaw Carpet, Prostrate ceanothus	6-9	⊕ ○	D to M
Cercis occidentalis	Western Redbud, California Redbud	5-9	⊕ ○	D to M
Cercis siliquastrum	Judas Tree, Redbud	6-9	⊕ ○	D to M
Cercocarpus montanus	Mountain Mahogany, Alderleaf mountain mahogany	6-7	○	D to M
Clitoria mariana	Atlantic Pigeonwings, Butterfly Pea	6-9	⊕ ○	D to M
Colutea arborescens	Bladder Senna	4-8	⊕ ○	D to M

Botanical Name	Common Name	USDA Hardiness Zones	Sun	Soil
Comptonia peregrina	Sweet Fern	3-6	⊕ ○	D to M
Cytisus decumbens	Prostrate Broom	5-8	○	D to M
Cytisus scoparius	Broom, Scotch broom, Common Broom	5-8	⊕ ○	D to M
Desmodium canadense	Showy tick-trefoil	3-6	⊕ ○	D to M
Desmodium glutinosum	Pointed-leaved Ticktrefoil	3-9	⊕ ○	M
Dryas octopetala	Mountain Avens	3-6	○	M
Elaeagnus angustifolia	Oleaster, Russian olive	2-7	○	D to M
Elaeagnus commutata	Silverberry	2-6	○	D to M
Elaeagnus glabra	Goat nipple	7-10	⊕ ○ ●	D to M
Elaeagnus multiflora	Goumi, Cherry silverberry	5-9	⊕ ○	D to M
Elaeagnus pungens	Elaeagnus, Thorny olive	6-10	⊕ ○ ●	D to M
Elaeagnus umbellata	Autumn Olive	3-7	○	D to M
Elaeagnus x ebbingei	Elaeagnus, Ebbing's Silverberry	5-9	⊕ ○ ●	D to M

Botanical Name	Common Name	USDA Hardiness Zones	Sun	Soil
Genista pilosa	Hairy greenweed, silkyleaf broom	5-8	⊕ ○	D to M
Genista pilosa procumbens	Creeping broom, Creeping hairy broom	6-8	⊕ ○	D to M
Genista sagittalis	Winged Broom, Arrow Broom	3-8	⊕ ○	D to M
Genista tinctoria	Dyer's Greenweed, Common Woadwaxen, Broom	4-7	○	D to M
Glycyrrhiza echinata	Wild Liquorice, Chinese licorice	7-10	⊕ ○	M
Glycyrrhiza glabra	Liquorice, Cultivated licorice	7-10	⊕ ○	M
Glycyrrhiza lepidota	American Liquorice	3-8	⊕ ○	M
Glycyrrhiza uralensis	Gan Cao	5-9	⊕ ○	M
Hedysarum boreale	Sweet Vetch, Utah sweetvetch, Northern sweetvetch	3-7	○	M
Hippophae rhamnoides	Sea Buckthorn, Seaberry	3-7	○	D,M,W
Hippophae salicifolia	Willow-Leaved Sea Buckthorn	3-7	○	D,M,W
Indigofera decora	Chinese indigo	5-7	○	M
Lathyrus japonicus maritimus	Beach Pea	3-7	○	D to M

Botanical Name	Common Name	USDA Hardiness Zones	Sun	Soil
Lathyrus latifolius	Perennial Sweet Pea, Perennial pea	5-9	⊕ ○	D to M
Lathyrus linifolius montanus	Bitter Vetch	5-9	⊕ ○	M
Lathyrus odoratus	Sweet Pea, Wild Pea, Vetchling	2-11	⊕ ○	M
Lathyrus tuberosus	Earthnut Pea, Tuberous sweetpea	5-9	⊕ ○	M
Lespedeza bicolor	Lespedeza, Shrub lespedeza	4-8	⊕ ○	D to M
Lespedeza capitata	Roundhead Lespedeza	4-8	○	M
Lespedeza maximowiczii		4-8	○	M
Lotus corniculatus	Bird's Foot Trefoil	3-8	○	D to M
Lotus uliginosus	Greater Bird's Foot Trefoil	5-9	○	M, W
Lupinus angustifolius	Blue Lupin, Narrowleaf lupine	7-9	○	M
Lupinus arboreus	Tree Lupin, Yellow bush lupine	7-10	○	D to M
Lupinus perennis	Sundial Lupine	4-8	○	D to M
Lupinus perennis	Sundial Lupine	4-8	○	D to M
Maackia amurensis	Chinese Yellow Wood, Amur maackia	4-7	○	D to M
Medicago sativa	Alfalfa, Yellow alfalfa	4-8	○	D to M

Botanical Name	Common Name	USDA Hardiness Zones	Sun	Soil
Myrica californica	Californian Bayberry	7-11	⊕ ○	M
Myrica cerifera	Bayberry Wild Cinnamon	7-11	⊕ ○	M
Myrica gale	Bog Myrtle, Sweetgale	2-9	⊕ ○	M, W
Myrica pensylvanica	Northern Bayberry	2-9	⊕ ○	D to M
Myrica rubra	Chinese Bayberry	9-11	⊕ ○	M
Onobrychis viciifolia	Sainfoin	5-9	○	M
Ononis spinosa	Spiny Rest Harrow	5-9	○	D to M
Phaseolus polystachios	Thicket Bean. Wild bean	6-10	⊕ ○	M
Prosopis glandulosa	Honeypod mesquite. Glandular mesquite	8-11	○	M
Psoralea esculenta	Breadroot, Large Indian breadroot	4-8	○	D to M
Robinia hispida	Bristly locust, Rose-acacia, or Moss locust	4-8	○	D to M
Robinia pseudoacacia	Black Locust, Yellow Locust	4-9	○	D to M
Robinia viscosa	Clammy Locust, Hartweg's locust	3-7	⊕ ○	D to M

Botanical Name	Common Name	USDA Hardiness Zones	Sun	Soil
Shepherdia argentea	Buffalo Berry, Silver Buffaloberry	3-9	⊕ ○	D to M
Shepherdia canadensis	Buffalo Berry	2-6	⊕ ○	D to M
Sophora japonica	Japanese Pagoda Tree, Scholar Tree	4-7	○	M
Strophostyles umbellata	Pink fuzzybean, Perennial wild bean	6-9	⊕ ○	D to M
Stylosanthes biflora	Sidebeak pencilflower	4-8	⊕ ○	D to M
Thermopsis villosa	Aaron's rod, Carolina lupine	5-8	⊕ ○	D to M
Trifolium incarnatum	Crimson Clover	3-9	○	M
Trifolium pratense	Red Clover	5-9	○	M
Trifolium repens	White Clover, Dutch Clover	4-8	○	M
Trifolium subterraneum	Subterranean Clover	7-10	○	M
Ulex europaeus	Gorse, Common gorse	5-9	○	D to M
Vicia americana	American Vetch, Mat vetch	4-7	⊕ ○	M
Vicia cracca	Tufted Vetch, Bird vetch, Cow vetch	4-8	⊕ ○	M

Botanical Name	Common Name	USDA Hardiness Zones	Sun	Soil
Vicia faba equina	Horsebean	4-8	⊕ ○	M
Vicia faba minuta	Tick Bean	4-8	⊕ ○	M
Vicia sativa	Winter Tares, Garden vetch, Subterranean vetch	4-8	⊕ ○	M
Wisteria floribunda	Japanese Wisteria	5-9	○	M
Wisteria frutescens	American Wisteria	4-8	○	M
Wisteria sinensis	Chinese Wisteria	5-9	○	M

GLOSSARY

Aggregates—particles of soil that have been bound together, for instance by fungal filaments, root hairs or sugars excreted by plant roots. Aggregation of soil is vital to its fertility and water-holding capacity.

Agroforestry—agriculture based on the food forest concept with multiple crops at different levels.

Allelopathy—a biological interaction between two organisms, sometimes negative (eg walnut trees exude a chemical which prevents other plants' growth) and sometimes beneficial.

Alley cropping—planting trees in rows and using the space between for other crops such as fruit bushes or annual vegetables.

Annual plant—a plant that completes its life cycle within a year, from germination to reproduction to death. Annuals are more energy demanding than perennials.

Berm—the bank on the downhill side of a swale.

Biennial plant—a plant which takes two years to go through its life cycle, bearing flowers and seed in its second year. Parsley is one example.

Biodiversity refers not just to having a large number of species, but having a large number of different relationships between those species. In permaculture, biodiversity ensures resilience, fertility and stability.

Biomass—biomass is the measure of organic matter, whether dead or alive. It represents the energy stored in a plant originating from photosynthesis. Biomass accumulation enhances fertility (eg a tree grows more leaves and that biomass can be used as mulch to improve the soil).

Catch crop—a fast-growing crop grown between sowings of main crops, in order to protect the soil. It can also be chopped and used as a green manure.

Chop'n'drop—the practice of chopping up plants in order to use them as a mulch where they grew.

Closed loop—also known as the 'circular economy' concept, a system in which everything is recycled and nothing is wasted.

Companion planting—putting plants together which have beneficial relationships with each other, such as support and climbing vine, nitrogen-hungry plant and nitrogen fixer, or a food plant and a pest repellent.

Compost is made by breaking down carbon- and nitrogen- containing organic materials and is used to improve fertility, nourishing the soil ecosystem.

Coppicing is the practice of periodically chopping down the trunks of trees such as hazel to the ground. It provides poles for use in the garden; the tree will regrow from the roots, usually forming several trunks in a circle and expanding over time.

Division is a means of propagating plants by digging up a clump or rhizome and dividing it into separate plants. Irises and lemon balm for instance need to be divided every few years to maintain their vigor.

Ecosystem—can describe the system as a whole, with its different energy flows and relationships, or the sum of the community of organisms that it shelters. Permaculture sees the gardener's task as managing an entire ecosystem.

Ecotone—a transition zone at the intersection of two different ecologies (eg wetland and grassland) which includes both the species that live in those two ecologies and species that are specially adapted to living in the ecotone. It is more biodiverse than either of the two bordering ecologies.

Edge—the concept of the 'edge' derives from the ecotone; edges offer advantages, such as partial shade at the edge of a tree's drip-zone, or living space for amphibians on the banks of a pool. Permaculture gardens are designed to maximize the use of edges.

Element—any plant, path, watercourse, building, etc within a permaculture. Each element is examined in terms of its functions and its relationship with other elements.

F1 hybrid—a hybrid seed bred for yield or pest-resistance. They are usually sterile, which means they cannot be used for seed saving.

Fertigation—a term coined for the use of greywater which includes nutrients, as a cross between irrigation and fertilizer.

Forest garden—a garden that imitates a mature forest, with mainly perennial plants growing in several layers. When focused on providing edible plants, known as a food forest.

Function—the way elements in the garden relate to each other. For instance a chicken's functions could be described as scratching the soil, eating weeds, making manure, laying eggs.

Germination—when a seed sprouts to form a seedling.

Green manure—often used to describe plants such as alfalfa and mustard which are good catch crops and can be cut down where they stand to provide nutrients for the soil.

Green mulch—plants chopped up to provide a mulch. These may be catch crops, or they may be the unwanted part of annual plants, such as tomato or squash vines.

Greywater—water that has been used in washing, bathing, or kitchen activities. It can be used on the garden, but cannot be stored unless it has been treated, eg in a reed bed.

Guerrilla gardening—sowing plants on derelict or wasteland sites, usually in an urban setting.

Guild—the concept of a small group of plants that form a satisfactory grouping, each element benefiting the others.

Herb spiral—a mound with a spiral path made on it by using rocks or other support. Herbs are growing in the microclimate best corresponding to their needs. For obvious reasons it's a good thing to put it close to your kitchen door!

Hugelkultur—'mound culture', made by piling different layers of organic material on top of tree trunks and branches in a pit. The wood rots down into humus very slowly and assists in water retention.

Humanure—manure coming from humans. Unless you are living off grid or use a composting toilet system, you really don't want to use this. (You may, however, pee on your compost heap; urine is a good compost activator).

Humus—organic matter that cannot decay further. It is humus that helps the soil retain water and a high percentage of humus makes soil more fertile.

Keyhole bed—a circular raised bed with a path into the center, an efficient way to grow plants.

Keypoint—the place where a slope transitions from convex to concave. The keyline is the contour line that links keypoints.

Leguminous plants—these are plants that are able to 'fix' nitrogen in the soil through their symbiotic relationship with nitrogen fixing bacteria. They include peas and beans, black locust trees and other plants. They are very useful in regenerating depleted soil.

Monoculture—the practice of growing a single crop extensively. It is the opposite of everything that permaculture is about and often depletes the soil.

Mulch—an organic material spread over the soil to insulate it, prevent the evaporation of water and eventually enrich the organic matter in the soil.

Mycelium—the vegetative part of a fungus, a network-like structure underneath the soil.

Needs and yields—analyzing elements in the garden by their requirements and what they produce. For instance a tomato plant needs soil, water, nitrogen and mulching, and produces tomatoes and green waste.

Nitrogen-fixer—see leguminous plants.

Organic agriculture—agriculture that does not use chemical inputs such as nitrate fertilizers or growth hormones. It is not necessarily a permaculture.

Organic material—can be made up of leaves, roots, flowers, waste food etc.

Organic matter—organic material that has been decomposed and incorporated into the soil.

Patterning—copying patterns that are found in nature.

Perennial plants—plants which live for more than two years, with a long cycle of growth. These are at the heart of permaculture and include trees, bushes, shrubs and perennial herbs.

Pioneer plant—a plant that has developed the ability to establish itself in depleted soil or in wasteland, by requiring little water or nutrients. Most pioneer plants are seen as 'weeds' by conventional gardeners.

Pollarding—a tree can be kept to a particular size or height by pollarding—cutting the branches back to the trunk on an annual or longer basis. Willows are often pollarded in order to harvest osiers for basket making.

Pollinator insect—an insect which crawls into flowers for food and in doing so transfers pollen from one flower to another.

Polyculture—agriculture in which different crops are grown together.

Propagation—the creation of new plants from the existing stock. Usually applied to making cuttings or root divisions.

Resilience—the ability to recover. Permaculture gardens are designed to have good resilience in the case of drought, flood, extreme temperatures, etc.

Rhizome—a horizontal underground plant stem or root stalk, such as that of the iris. They can be divided in order to propagate new plants.

Sector—any energy entering the garden, such as sunlight, a stream, wind, or pollution.

Seed saving—saving the seeds from (usually annual) plants that have been grown in the garden, instead of buying a new packet.

Self-fertile—fruit trees which can pollinate themselves. Some fruit trees need the presence of another species as pollinator.

Sheet mulch—mulch laid down on an extensive area, usually as a first step in recovering depleted or weedy ground.

Silvopasture—agriculture which integrates forestry, forage plants and grazing for domesticated animals such as goats or sheep.

Succession—the natural process by which an ecosystem grows from a few pioneer plants into a full scale forest.

Swale—a dry ditch dug along a contour line to enable water to be absorbed at that point.

Taproot—a long, thin root that penetrates deeply into the soil.

Three Sisters—a companion planting of corn, beans and squash that was (and is) extensively used by Native American peoples. Sometimes the Four Sisters, along with Rocky Mountain Bee Plant.

'Weed'—a plant that is growing where you don't want it. Permaculture sees many 'weeds' as useful plants, for instance in making a green mulch or opening up the soil with their roots.

Wormery—a composting bin for food scraps and other kitchen waste which uses Red

Wrigglers or other specialized worm species to convert the waste to compost.

Zone—permaculture gardens are divided into zones, depending on the type of management of the zone (for instance, Zone 1 is intensively cultivated while Zone 5 is a rarely maintained wilderness Zone).

BIBLIOGRAPHY

WEBSITES

www.pfaf.org
Large database of many plants, their usage, soil type and much more

www.rhs.org.uk
Large database of plants and tips on growing

www.permaculture.org.uk
Basics of permaculture and offered courses

www.planetnatural.com
Organic gardening and pest control advice, research center

www.piwakawakavalley.co.nz
A family homestead providing some useful information

www.treehugger.com
Modern sustainability site that offers advice, clarity, and inspiration for both the eco-savvy and the green-living novice

www.gardenorganic.org.uk
Useful information on organic gardening

www.gardenerspath.com
Useful information on many things such as animals and wildlife and enriching the soil

www.gardeningknowhow.com
Useful information on all things gardening

www.houstonchronicle.com

www.neverendingfood.org
Website dedicated to Permaculture and nutrition in Malawi, Africa

www.gardenia.net
A useful plant finder resource plus much more

www.feis-crs.org
References on the distribution, biology, ecology, and fire responses of organisms in North America

https://biocontrol.entomology.cornell.edu/weedfeedTOC.php
Biological control information

www.shootgardening.co.uk
Plant search, useful articles and forums

www.gardenersworld.com
UK's best known gardening resource

www.davesgarden.com
Plant search, useful articles and forums

www.practicalplants.org
Plant encyclopaedia

www.woodlandtrust.org.uk
Promoting woodland and tree planting in the UK

https://deepgreenpermaculture.com/
Australian, very practical site by a permaculturist who trained with Holmgren and Mollison

https://www.gardeninthehimalayas.org/
Inspiring project in the heart of Nepal that preserves native plants, inspires local communities and protects the natural habitat

https://www.pandorathomas.com/epc
Earthseed Permaculture Center, Sonoma's first All Black owned and run Permaculture farm

https://permacultureapprentice.com/
Roadmap from zero to permie hero in five years, with some excellently illustrated guides and free training series

https://www.permaculturenews.org/
Run by Geoff and Nadia Lawton. A not-for-profit organization where you can find a lot of useful information on permaculture plus permaculture courses. They have a permaculture demonstration site in Channon, NSW, Australia

https://www.permaculture.co.uk/
Abundance of useful information, videos, podcasts, forums

https://www.tenthacrefarm.com/
With a focus on suburban permaculture and homesteading

https://treeyopermacultureedu.com/
Kentucky based, highly informative site including a Pawpaw Masterclass

www.beebuilt.com
Guide to bee keeping, plus supplies and courses

https://projects.sare.org/sare_project/fnc14-944/
Pasture weed control with juglone

https://agsci.source.colostate.edu/as-a-way-to-fight-climate-change-not-all-soils-are-created-equal
Study on soil

https://www.kansaspermaculture.org/post/4-benefits-of-contour-farming
Nebraska study on contour farming

*https://www.researchgate.net/
publication/303924502_Effect_of_Plant_
Shading_and_Water_Consumption_on_
Heat_Reduction_of_Ambient_Air*
Maejo University study—Effect of Plant Shading and Water Consumption on Heat Reduction of Ambient Air

*http://puyallup.wsu.edu/wp-content/uploads/
sites/403/2015/03/companion-plantings.pdf*
This paper is suspicious of the 'myth of companion plantings'—but points out that plant associations, particularly at root level and through mycorrhiza, *do* have an influence on growth

www.papergardenworkshop.com
Bubble diagrams

*https://sivanandayogafarm.org/
what-are-the-permaculture-zones/*
Garden zones

www.jardin-reve.fr
Garden zones

www.livingpermaculturepnw.com
Permaculture information

https://permaculturefoodforest.wordpress.com
Permaculture information

www.midwestpermaculture.com
Permaculture design and courses

www.waldeneffect.org
Blog

www.oasisdesign.net
Water system designs

*https://www.letsgogreen.com/
greywater-recycling.html*
Greywater recycling

www.bctribune.com
Rainwater system design

www.gardeners.com
Gardening advice and supplies

www.offgridpermaculture.com
Useful information about living off grid

*https://www.fix.com/blog/
raising-chickens-at-home/*
Chicken tractor example

www.finegardening.com
Lots of useful gardening information

www.gardeninminutes.com

BOOKS

Bane, Peter '*The Permaculture Handbook: Garden Farming for Town and Country*' 2012

Crawford, Martin '*Creating a Forest Garden*' 2010

Fukuoka, Masanobu '*The One-Straw Revolution: an Introduction to Natural Farming*' 1975

Hart, Robert A de J '*Forest Gardening: Cultivating an Edible Landscape*' 1996

Hemenway, Toby '*Gaia's Garden: A Guide to Home-Scale Permaculture*' Second edition, 2009

Kemp, Juliet '*Permaculture in Pots: How to Grow Food in Small Urban Spaces*' 2013

Kourik, Robert '*Designing and Maintaining Your Edible Landscape Naturally*' 2014

Mollison, Bill '*Permaculture- A Designers' Manual*' 1988

Whitefield, Patrick '*How to Make a Forest Garden*' 1996

Hewitson Best, Robin, Ward, J.T, '*The Garden Controversy: A Critical Analysis of the Evidence and Arguments Relating of the Production of Food from Gardens and Farmland*'—University of London, 1956

YOUTUBE AND YOUTUBE CHANNELS

https://www.youtube.com/channel/UCL_r1ELEvAuN0peKUxI0Umw
Discover Permaculture with Geoff Lawton

https://www.youtube.com/channel/UC24CMfYNfyr5tAq_AZB03MA
John Trevethen—permaculture homestead on San Juan island

https://www.youtube.com/channel/UChqxlr587JCD5I9veE62s1w
Verge Permaculture

https://www.youtube.com/watch?v=0uBz0yzRTTE
Getting started in beekeeping—Norfolk Honey Company

RESOURCES

ONLINE PLANT DATABASES

www.pfaf.org
An abundance of information of many plant species, their soil type, uses and much more information

USDA plants database
https://plants.usda.gov/home
Search for plants, the hardiness zone and a lot of other useful information

www.ars-grin.gov
USDA's Germplasm Resources Information Network database (GRIN) where you can find information on useful plants.

PERMACULTURE MAGAZINES

Permaculture, UK
www.permaculture.co.uk
Permaculture and practical solutions

North American *Permaculture Magazine*
www.permaculture,ag.org
Offshoot of Permaculture Magazine, UK for Canada, Mexico, USA

PIP magazine
www.pipmagazine.com.au
Permaculture and sustainable living

Permaculture Design Magazine
www.permaculturedesignmagazine.com
For the regeneration of human habitat

TEACHING AND CONSULTING—PERMACULTURE

Central Rocky Mountain Permaculture Institute, Basalt Colorado
www.crmpi.org

Earthflow Design Works, San Luis Obispo, California
www.earthflow.com

The Farm Ecovillage Training Center, Summertown, Tennessee
www.thefarm.org

Finger Lakes Permaculture Institute, Ithaca, New York
www.fingerlakespermaculture.org

Lost Valley Educational Center,
Dexter, Oregon
www.lostvalley.org

Occidental Arts and Ecology Center,
Occidental, California
www.oaec.org

Regenerative Design Institute, Bolinas,
California
www.regenerativedesign.org

Regenerative (Online Courses)
www.regenerative.com

Yester Morrow, Waitsfield, Vermont
www.yestermorrow.org

Permaculture Institute, UK
www.permaculture.org.uk
A range of online courses

Free Permaculture
www.freepermaculture.com
Free online courses

School of permaculture, Texas
www.schoolofpermaculture.com

Permaculture Academy, Los Angeles,
California
www.permacultureacademy.com

Midwest Permaculture, Youngstown, Ohio
www.midwestpermaculture.com

Terra Alta, Sintra, Portugal
www.terralta.org

Earthseed Permaculture Center and Farm,
Sonoma County, California
www.pandorathomas.com

Permaculture Research Institute, Channon,
New South Wales, Australia
www.permaculturenews.org
Run by Geoff and Nadia Lawton

GARDEN SUPPLIES AND PLANTS, SEEDS

Banana Tree, Easton, Pennsylvania
www.banana-tree.com
Tropical seeds and bulbs

Open Circle Seeds, Medocino County,
California
www.opencircleseeds.com
Organic seeds

Redwood Seeds, North central California
www.redwoodseeds.net
Organic seeds

Resilient Seeds, Ferndale, Washington
www.resilientseeds.com
High quality seeds

Wild Garden Seed, Philomath, Oregon
www.wildgardenseed.com
Seeds and flowers

Adaptive Seeds, Sweet Home, Oregon
www.adaptiveseeds.com
Farm-based company specializing in seeds for the NW climate

Native Seeds Search, Tucson, Arizona
www.nativeseeds.org
Ancient native varieties of seeds

Prairie Road Organic Seed, Fullerton, North Dakota
www.prairieroadorganic.co

Seed Savers Exchange, Decorah, Iowa
www.seedsavers.org
Farm-based company

Southern Exposure Seed Exchange, Mineral, Virginia
www.southernexposure.com
Well chosen seeds for Southern weather

Fedco Seeds, Clinton, Maine
www.fedcoseeds.com
Cooperative, broad choice of seeds

Strictly Medicinal Seeds, Oregon
www.strictlymedicinalseeds.com
Seed and growing advice

The Cook's Garden, Hodges, South Carolina
www.cooksgarden.com
Retail and wholesale seeds

Deep Diversity, Gila, New Mexico
www.one-garden.org
Seeds

Edible Landscaping, Afton, Virginia
www.eat-it.com
Wide selection of fruit, vegetables, shrubs

Forestfarm Nursery, Williams, Oregon
www.forestfarm.com
Large variety of useful plants

Harmony Farm Supply and Nursery, Craton, California
www.harmonyfarm.com
Seeds and supplies

Hidden Springs Nursery, Cookeville, Tennessee
www.hiddenspringsnursery.com
Unusual fruits and other plants

One Green World, Molalla, Oregon
www.onegreenworld.com
Large selection of plants, trees and more

Ornamental Edibles, San Jose, California
www.ornamentaledibles.com
Gourmet seeds and edible flowers

Peaceful Valley Farm and Garden, Grass Valley, California
www.groworganic.com
Seeds, irrigation and more

Plants of the Southwest, Santa Fe, New Mexico
www.plantsofthesouthwest.com
Native plants & seeds from the SW

High Mowing Seeds, Wolcott, Vermont
www.highmowingseeds.com

The Organic Gardening Catalogue, UK
www.organiccatalogue.com
A wide range of organic seeds, flowers and gardening equipment

Suttons, UK
www.suttons.co.uk
A wide range of organic seeds, flowers and gardening equipment

Ethical Organic Seeds, UK
www.ethicalorganicseeds.co.uk
Organic seeds, company based on ethical principles

BOOKS

'Gaia's Garden: A Guide to Home-Scale Permaculture—2nd Edition'—
Toby Hemenway
2009

'Sepp Holzer's Permaculture: A Practical Guide for Farmers, Smallholders and Gardeners: 1'—Sepp Holzer
2010

'The Nature and Properties of Soils (14th Edition)'—Nyle C Brady
2013

'Breed Your Own Vegetable Varieties: The Gardener's and Farmer's Guide to Plant Breeding and Seed'—Carol Deppe
2000

'Introduction to Permaculture'—Bill Mollison and Reny Mia Slay
2013

'Farming the Woods: An Integrated Permaculture Approach to Growing Food and Medicinals in Temperate Forests'—Ken Mudge and Steve Gabriel
2014

'Integrated Forest Gardening: The Complete Guide to Polycultures and Plant Guilds in Permaculture Systems'—Wayne Weiseman
2014

'Hepburn Permaculture Gardens; 10 Years of Sustainable Living'—David Holmgren
2001

'Natural Companions: The Garden Lover's Guide to Plant Combinations'—Ken Druse
2012

'Four-season Harvest: Organic Vegetables from Your Home Garden All Year Long, 2nd Edition'—Eliot Coleman
2005

'Raising Chickens: Beginners Guide to Raising Healthy and Happy Backyard Chickens'—Janet Wilson
2020

'Permaculture: A Designer's Manual'—Bill Mollison
1988

'Backyard Farming on an Acre (more or less)'—Angela England
2012

'Designing and Maintaining Your Edible Landscape—Naturally'—Robert Kourik
2004

'Stress-Free Chicken Tractor Plans: An Easy to Follow, Step-by-Step Guide to Building Your Own Chicken Tractors'—John Suscovich
2016

'Rainwater Harvesting for Drylands and Beyond, Volume 1: Guiding Principles to Welcome Rain Into Your Life and Landscape'—Brad Lancaster
2019

'Bioshelter Market Garden: A Permaculture Farm'—Darrell Frey
2010

'Tree Crops: A Permanent Agriculture'—Russell Smith
2018

'Weedless Gardening'—Lee Reich
2001

'Earth Users Guide to Permaculture'—Rosemary Morros
2007

'Perennial Vegetables & Perennial Vegetable Gardening'—Eric Toensmeier
2012

'The Complete Idiots Guide to Composting'—Chris McLaughlin
2010

'How to Make a Forest Garden'—Patrick Whitefield
2012

'The Earth Care Manual: A Permaculture Handbook for Britain and Other Temperate Climates'—Patrick Whitefield
2016

'Forest Gardening'—Robert Hart
1996

'Home Landscaping (Series)'—
Roger Holmes
(series on different regions in the US, covering landscaping basics and suitable regional plants)

'Water for Every Farm: Using the Keyline Plan'—P. Yeomans
2021

'Gardening Without Work: For the Aging, the Busy & the Indolent'—Ruth Stout and Steven Stiler
2021

'Homegrown Herbs: A Complete Guide to Growing, Using and Enjoying More than 100 Herbs'—Tammi Hartung
2015

'The One-Straw Revolution: An Introduction to Natural Farming'—Masanobu Fukuoka
2020

'Lasagna Gardening: A New System for Great Gardens: No Digging, No Tilling, No Weeding, No Kidding!'—Patricia Lanza
1999

'Uncommon Fruits for Every Garden'—Lee Reich
2004

'Edible Forest Gardens'—David Jacke and Eric Toensmeier
2006

'The Backyard Homestead: Produce all the Food You Need on just a Quarter Acre!'—Carleen Madigan
2009

'How to Attract Birds to Your Garden: Foods they like, plants they love, shelter they need'—Dan Rouse
2020

'Teaming with Microbes: The Organic Gardener's Guide to the Soil Food Web'—Jeff Lowenfels and Wayne Lewis
2010

'The Permaculture Handbook; Garden Farming for Town and Country'—Peter Bane and David Holmgren
2012

'*The Resilient Farm and Homestead: An Innovative Permaculture & Whole Systems Design Approach*'—Ben Falk
2013

'*Weeds and What They Tell Us*'—Ehrenfried E. Pfeiffer
2012

'*The Urban Homestead: Your Guide to Self-Sufficient Living in the Heart of the City*'—Kelly Coyne and Erik Knutzen
2010

Made in the USA
Middletown, DE
10 September 2024

60728680R00223